SAS® Graphics for Clinical Trials by Example

Kriss Harris
Richann Watson

sas.com/books

Contents

About This Book

What Does This Book Cover?

As a programmer working in the health care and life science industries, you are sometimes required to create graphs of the clinical trial results. The graphs are typically used by statisticians in the review of the data, as well as for regulatory submissions. This book is intended to help illustrate how to create these various graphs. Although there are other books out there that show how to use the **Statistical Graphics (SG) procedures and Graph Template Language (GTL)**, it is sometimes hard to translate that information into real-world use within the clinical industry. This book demonstrates through the use of examples, how to create some simple and straightforward graphs, as well as some complex graphs using CDISC ADaM data sets as the source in many examples. This book provides an introduction to the Output Delivery System (ODS) as well as a high-level overview of the basics of several SG procedures and GTL. This book does not go into all the SG procedures or all plots statements and options available. To cover all SG procedures, plots and available options are beyond the scope of this book.

Is This Book for You?

This book is designed to aid programmers working in the pharmaceutical industry to create graphs. Regardless if you are just learning how to produce graphs or have been working on graphs for a while. If you want to gain a better understanding of the reason why you are asked to produce graphs for clinical trials and want to know how to create specific types of graphs used for clinical trials, then this book is aimed at users like you.

What Are the Prerequisites for This Book?

In order to produce some of the graphs in this book, the data that you are working with might need to be manipulated so that it is in a format that can be used. Therefore, an understanding of Base SAS procedures and the DATA step is beneficial. Although not required, it can be useful to have a basic understanding of the SG procedures and GTL. Knowing how to create macro variables and macros would be advantageous.

What Should You Know about the Examples?

This book includes tutorials for you to follow and gain hands-on experience with SAS.

Software Used to Develop the Book's Content

All programs used to generate the graphs illustrated in this book are developed using SAS version 9.4 maintenance release 5.

Example Code and Data

The data used in this book is a combination of data from the CDISC pilot study, summary data sets based on the CDISC pilot study, and data created by the authors. The data from the CDISC pilot can be downloaded from the Updated Version of Pilot Submission Package (2013) https://www.cdisc.org/sdtmadam-pilot-project.

The data in Chapter 5 is fictitious data created by the authors. The authors are not oncology experts; however, they have worked on oncology studies where they had to produce time-to-event (TTE) graphs. Brief background information about TTE endpoints and types of graphs used are included to provide the reader with a high-level understanding of the type of data and output.

Also, in Chapter 6, the data is fictitious data created by the authors. The authors are not RECIST experts nor oncology experts. However, they have worked with both types of data and give a brief description and background on some concepts to help illustrate the type of outputs being created.

Chapter 8 utilizes a data set that contains two patients. Data for patient ABC-DEF-0001 is data based on one of the author's dogs who had an autoimmune disease, diabetes, and bladder cancer. The author had to vigilantly monitor blood work, events, and medications. Data for patient ABC-DEF-0002 is fictitious data, and it is not used in any of the examples.

The use of the word "patients" or "subjects" is typically determined based on the client. For this book, we opted to use the word "patients." However, there are some SAS procedures that produce default graphs that have the word "subjects" in a label.

Although all data sets are provided to help with executing the programs, they are not necessarily following standard CDISC ADaM data set naming conventions. Typically, these data sets would be temporary data sets created at time of graph creation. They are saved as permanent data sets to facilitate the execution of the programs that are found in the book.

CustomSapphire is a custom style template used specifically for the book, but was based on one of the standard SAS style templates - Sapphire. However, any of the standard SAS style templates could have been used.

All the data and code used in this book are available at https://github.com/rwatson724/SAS-Graphs-Clinical-Trials-Example.

You can access the example code and data for this book by linking to its author page at https://support.sas.com/authors.

Output and Graphics

All graphs shown in this book are created using the programs described in each chapter. However, in some instances, the full code is not available in the book in order to save space. The full code for each graph can be found https://github.com/rwatson724/SAS-Graphs-Clinical-Trials-Example.

We Want to Hear from You

SAS Press books are written *by* SAS Users *for* SAS Users. We welcome your participation in their development and your feedback on SAS Press books that you are using. Please visit sas.com/books to do the following:

- Sign up to review a book
- Recommend a topic
- Request information on how to become a SAS Press author
- Provide feedback on a book

Do you have questions about a SAS Press book that you are reading? Contact the author through saspress@sas.com or https://support.sas.com/author_feedback.

SAS has many resources to help you find answers and expand your knowledge. If you need additional help, see our list of resources: sas.com/books.

About These Authors

Kriss Harris worked at GlaxoSmithKline as a statistician supporting drug discovery. At GSK, he developed an increasing passion for both SAS graphics and teaching and taught SAS graphics to SAS programmers, statisticians, and scientists. After leaving GSK, he became an independent statistical programmer and has consulted at Eli Lilly, Eisai, and MedAvante-ProPhase. Currently, Kriss is consulting at Eli Lilly supporting drug reimbursements within the oncology therapeutic area and at MedAvante-ProPhase supporting the construction of edit checks. Kriss is based in London, England, and holds a bachelor's degree in statistics and internet computing from Kingston University and a master's degree in statistics from the University of Sheffield.

https://www.linkedin.com/in/krissharris/
https://krissharris.co.uk/

Richann Watson is an independent statistical programmer and CDISC consultant based in Ohio. She has been using SAS since 1996 with most of her experience being in the life sciences industry. She specializes in analyzing clinical trial data and implementing CDISC standards. Additionally, she is a member of the CDISC ADaM team and various sub-teams.

Richann loves to code and is an active participant and leader in the SAS user group community. She has presented numerous papers, posters, and training seminars at SAS Global Forum, PharmaSUG, and various regional and local SAS user group meetings. Richann holds a bachelor's degree in mathematics and computer science from Northern Kentucky University and master's degree in statistics from Miami University.

https://www.linkedin.com/in/richann-watson-31435422/
https://www.linkedin.com/company/28145660/admin/
https://datarichconsulting.com/

Learn more about these authors by visiting their author pages, where you can download free book excerpts, access example code and data, read the latest reviews, get updates, and more:

http://support.sas.com/harris
http://support.sas.com/watson

Acknowledgments

Acknowledgments

This book incorporates the efforts of many reviewers of the draft chapters. The SAS technical consultants took time from their busy schedules to test our programs and provide critical feedback. We are incredibly grateful to Debpriya Sarkar, Jane Eslinger, Lingxiao Li, and Prashant Hebber. Thank you also to Suzanne Morgen, Catherine Connolly, Robert Harris, and Denise Jones at SAS Press.

Kriss Harris' Acknowledgments

I would like to extend my sincere thanks to Sharon Carroll for allowing me to have the space and time to work on the book by taking the "lion" share of responsibility with watching over our children.

I am also grateful to 李菊 (Ju Li) for her unwavering emotional support, especially in times of need, and for helping me work on my purpose and complete the book.

I gratefully acknowledge the assistance of Victoria Lambrianidi and Kristina Liu for providing practical suggestions and for helping to coach me with completing the book.

Richann Watson's Acknowledgments

I would like to thank my husband, Tyler Watson, for his support, both emotionally and mentally during this process. He was my rock during the development of the book and took on the brunt of the household duties to help me achieve this goal.

In addition, I want to thank Tyler for allowing me to share our story about Chewy.

I am also grateful to my friends and colleagues: Louise Hadden, Deanna Schreiber-Gregory, Charu Shankar, Josh Horstman, and Troy Hughes, who provided help with review of content that was referenced in this book and who were there for me emotionally when I felt that I was at my wits' end.

Chewy's Story

When we got Chewy on February 13, 2004, we did not fathom how much of an impact he would have on our lives. In the first few weeks that we had Chewy, we almost lost him. He was very ill because the people that we got him from did not take care of him properly; he was riddled with parasites. The vet also discovered he had a dislocated leg, and it had apparently been that way since birth. They set his leg and we were free to take the little guy home.

Chewy loved doing things with us. He would go to car shows, boating, tubing behind the boat, ATVing, and kayaking. You name it, and he would do it. He was determined to join us on our outdoor adventures. He was so full of life.

Time flew by and after about 8 years, we started to notice that Chewy was sleepy for several days after any activity, and began limping around, even if the activity was as low-key as sitting outside at a car show. The vet put him on steroids, and he was back to normal (or so we thought). But soon, my husband noticed Chewy becoming more and more lethargic. After running many tests, a specialty vet determined that he had an autoimmune condition. We then began the arduous search for therapy. The only thing that seemed to work consistently was steroids. The vet wanted to see whether we could wean him off the steroids due to the risk of long-term side effects, but he always managed to backslide. We eventually found a dosage that he could live on for the remainder of his life. Things were great for almost another 4 years. Chewy continued to exhibit such a zest for life; he lived for being with us, wherever we were and whatever we were doing.

In December 2015, things turned for the worse again. At first we thought it was because his routine had been disrupted with having out-of-town family over for the holidays, but once they left, we noticed he was not bouncing back. He also started to have accidents in the house. Once again we rushed to the family vet. By this time, we had access to the family vet via messenger and text message, and they were always gracious to help us in a pinch. They relayed to us that Chewy was now diagnosed with diabetes. His glucose count was almost 700 mg/dL, which was close to being fatal! They gave him a relatively high dose of Vetsulin (pet version of insulin) for his weight. We monitored him closely and he did fine at the high dose, but then his glucose started creeping back up. We could not understand why, but then we realized his steroids and insulin were competing against each. Chewy was eventually switched to human insulin and it became a juggling act with the vets and us working together to manage his meds and blood counts. Most dogs would have freaked and panicked about going to the vets so frequently, but Chewy sat patiently at the vets and would greet his caregivers with tail wags and sometimes kisses.

Things were going well again for a while. We had his autoimmune condition and diabetes under control. My husband and I learned how to test his glucose levels at home. But then in December 2016, he started having accidents again and begging to go outside every 10 minutes. Our immediate thought was his glucose is creeping back up, but it was not. Chewy was eventually diagnosed with bladder cancer, which was inoperable. We could not understand why this sweet dog with such a love for life and people kept suffering from these ailments. The vet put him on chemotherapy, and he was fine for a while. A year into his chemotherapy treatment, the vet did some x-rays to see whether the tumor was shrinking. Unfortunately, it was not shrinking, so they decided to do a more aggressive treatment. Typically, it takes 2 to 3 vet techs to hold the dog down and keep the dog calm while they do the slow drip chemotherapy because the dogs are very nervous, but Chewy was calm and relaxed. This little guy wanted to live! He still wanted to go on outdoor adventures with us.

Two months after starting the slow drip chemotherapy, Chewy herniated several discs in his back and neck and became paralyzed. He looked at us and we could tell he was done fighting. Chewy drew his last breath on January 19, 2018, surrounded by people that loved and adored him. He touched many lives, not only in our family but in the SAS community. He loved wearing his green polo with his "I love SAS users".

The reason I share this story is because although nearly all the data in this book is fictitious data, the data in Chapter 8 is not. Patient 0001 is Chewy's data. When he got sick, I started tracking all his data just to help monitor him, and the programmer in me needed to see whether there was a pattern. I needed to see visually how all the data was working together, or not. Chewy and his data led me to discover how SAS GTL could help me, and others, visualize essential trends.

Chapter 1: Introduction to Clinical Graphics

1.1 Introduction to Output Delivery System Statistical Graphics

There are several procedures in SAS that use the Output Delivery System (ODS) Statistical Graphics software. Part of the ODS Graphics are Statistical Graphics (SG) procedures and Graph Template Language (GTL). SG procedures generate plots that are based on procedure-driven GTL templates. In other words, the templates are determined based on the plot statements and options provided within the SG procedure. In addition, you can generate graphs from several analytical procedures, such as the LIFETEST procedure. With ODS Graphics, you are able to obtain graphs automatically through the analytical procedures or create custom graphs using SG procedures or GTL (Matange, 2016).

Knowing ODS Graphics enables you to modify or add features that are based on the procedure-driven templates as well as create customized graphs that might not otherwise be possible. GTL makes it easier to incorporate features, such as embedding a table of data within the output area or displaying multiple graphs, regardless of whether the graphs are of the same or different type on the same page. Prior to GTL, these features might have been difficult to achieve. Furthermore, according to Matange (2013) some additional reasons to learn GTL are:

- GTL provides in one system the full set of features that you need to create graphs from the simplest scatter plots to complex diagnostics panels.

- GTL is the language used to create the templates shipped by SAS for the creation of the automatic graphs from the analytical procedures. To customize one of these graphs, you will need to understand GTL.

- GTL represents the future for analytical graphics in SAS. New features are being added to GTL with every SAS release.

The following conventions are used in the text (and not the SAS code) within this book. Terms that are in all capital letters denote SAS procedures, SAS statements, data sets, and variables in a data set (for example, PROC SGPLOT, the SCATTER statement, the ADAE data set, and the

AESTDT variable). Terms that are in italic capital letters denote macro variables (for example, *SYSDATE* and *SYSTIME).* Bold italic letters denote macros (for example, **%SurvivalTemplateRestore**).

1.1.1 Understanding the Basic: Statistical Graphics Procedures

SG procedures can be used to create a number of graphs. They use clear and concise syntax. SG procedures include these two procedures: SGPLOT and SGPANEL. Although there are other SG procedures, SGPLOT and SGPANEL are the only SG procedures that are used in the production of some of the graphs illustrated in the book. The SGRENDER procedure is used along with the TEMPLATE procedure to produce custom graphics. These two procedures combined are referred to as GTL in this book.

1.1.1.1 Overview of the SGPLOT Procedure

For many graphs, the SGPLOT procedure is typically sufficient. In order to use SGPLOT at least one plot statement must be specified. Within SGPLOT you can overlay different plots so that they occupy the same data area in the graph. Types of graphs that can be produced with SGPLOT include bar charts, high-low plots, block plots, series plots, and waterfall plots.

Although only one plot statement is needed, further control over the graph's appearance can be achieved with the use of additional statements. You can control various aspects such as colors and markers that are associated with a style with the use of STYLEATTRS. If there is a particular Unicode character or an image that you want to use in one or more plots specified within the SGPLOT procedure, you can define the marker symbol using the SYMBOLCHAR or SYMBOLIMAGE. In addition, you can control the axes options with the use of XAXIS, X2AXIS, YAXIS, and Y2AXIS. XAXIS and X2AXIS control the X (bottom X axis) and X2 (top X axis) respectively; while YAXIS and Y2AXIS control the Y (left Y axis) and Y2 (right Y axis) respectively.

Each statement within SGPLOT has additional options that allow for further control of the appearance. The options vary based on the plot specified, but you can change other attributes like color, text rotation, fill pattern, and line pattern. The syntax for SGPLOT is simple as illustrated in Program 1-1.

Program 1-1: SGPLOT Procedure Syntax

```
proc sgplot <options>;
   <... plot statements ...> / <options>;
run;
```

1.1.1.2 Overview of the SGPANEL Procedure

With the SGPANEL procedure, you can create a panel of similar graphs based on one or more classification variables. In order to use SGPANEL, the PANELBY statement must be specified. The PANELBY statement determines the different combinations of the classification variables. At least one classification variable needs to be provided when using SGPANEL. With SGPANEL, SAS automatically determines the panel layout, that is, SAS determines the number of individual cells for a data area in the graph, and, if necessary, SAS creates multiple graphs if the number of unique combinations of classification variables does not fit on one graph.

Other than the use of PANELBY, SGPANEL can use the same plot statements and options to help control the appearance of each individual cell in the graph. Program 1-2 demonstrates the standard syntax for SGPANEL.

Program 1-2: SGPANEL Procedure Syntax

```
proc sgpanel <options>;
   panelby variable / <options>;
   <... plot statements ...> / <options>;
run;
```

1.1.1.3 Overview of the SGSCATTER Procedure

While the SGSCATTER procedure can be used to create a simple scatter plot of Y by X, it can also be used to create a panel of scatter plots that are based on the different combinations of variables specified. The type of panel created is determined based on whether the PLOT, COMPARE, or MATRIX statement is used. The use of PLOT, COMPARE, or MATRIX statement is required when using SGSCATTER. In addition, each one of the statements have their own set of options that enable you to control the appearance of the plots being produced.

To produce a simple scatter plot similar to what would be produced if you used SGPLOT with the SCATTER plot statement, you would use the PLOT statement and specify your Y and X variables. The PLOT statement creates a cell for each combination of Y variables and X variables. For example, if there are two Y variables and one X variable specified, then two cells are created – one for the first Y variable by the X variable and another for the second Y variable by the X variable.

With the use of the COMPARE statement, you can achieve similar results as with the PLOT statement, with the main difference being that the COMPARE statement produces a panel with shared axes, while the PLOT statement has individual axes for each X and Y variable combination.

The MATRIX statement produces a matrix of scatter plots. The number of cells within the matrix is determined by the number of numeric variables that are specified. The downward diagonal from left to right contains the variable name or label.

When specifying more than one X or Y variable, the variables should be enclosed in parenthesis and should be separated by a space. Refer to Program 1-3 for general syntax for SGSCATTER for PLOT, COMPARE, and MATRIX.

Program 1-3: SGSCATTER Procedure Syntax

```
proc sgscatter <options>;
   plot ( y-variable(s) ) * ( x-variable(s) );
run;

proc sgscatter <options>;
   compare x = ( x-variable(s) )
           y = ( y-variable(s) ) / <options>;
run;

proc sgscatter <options>;
   matrix numeric-variable(s) /<options>;
run;
```

1.1.2 Understanding the Basics: Graph Template Language

While SG procedures do produce stellar graphs, there are times when you might want to customize specific aspects of the graph or you might need to produce a graph that is not easily done using the SG procedures.

In order to use GTL, it is important that you first understand the basics. This includes understanding the difference between the graphics output area and the procedures output area. In addition, you should have an understanding of the difference between single-cell and multi-cell graphs and the various layouts associated with each. You should also have knowledge of the basic syntax for the most common layouts.

1.1.2.1 Draw Spaces for a Graph

Before you start using GTL, you need to know the difference between the types of drawing space in a graph. There are four different drawing spaces: data, wall, layout, and graph.

The data area is the innermost portion of the output as shown in Figure 1-1. This is the portion of the graph that contains the data, that is, the portion where the plots are displayed. While most of the graphs produced have a data area, there are a few that do not. "The data area does not apply to graphs that do not have axes, such as pie charts" (SAS Institute Inc., 2016). Graphs without axes, such as pie charts, are drawn using a REGION layout.

Figure 1-1: Data Area of Graph

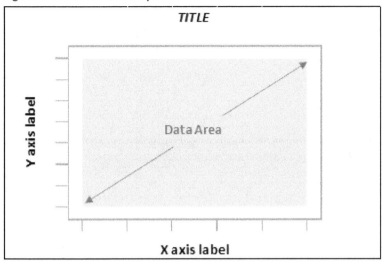

Notice how the data area does not include the space up to the axes. This space is part of the wall area, which also encompasses the data area, as illustrated in Figure 1-2. The wall area is bound by the X and Y axes and X2 and Y2 axes if they are used. Similar to the data area, if there are no axes, then there is no wall area.

Figure 1-2: Wall Area of Graph

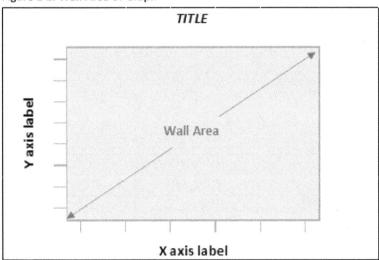

The part of the graph that includes additional information such as axis-labels, tick marks and legends is in the layout area. Any information that is displayed on the graph that is defined within a LAYOUT block is part of the layout area as seen in Figure 1-3.

Figure 1-3: Layout Area of Graph

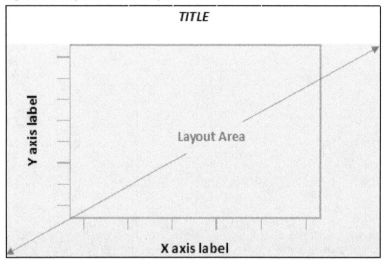

The final area is the graph area, and this encompasses the entire graph as shown in Figure 1-4.

Figure 1-4: Graph Area of Graph

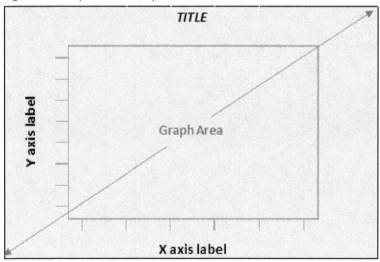

Figures 1-1 through 1-4 illustrate the various areas for a single-cell graph. When the plots occupy the entire data area, whether there is a single plot or multiple plots that are overlaid (for example, multiple GTL plot statements within an overlay so that the plots appear on top of each other within the same data area), this is referred to as a single-cell graph. When each plot occupies a different data area, this is known as a multi-cell graph. Multi-cell graphs can either be pre-defined or data-driven. A pre-defined multi-cell graph breaks the graph area into pre-defined portions so that each portion represents different pieces of information that include the data, axes, headers, and titles. The data-driven multi-cell graph breaks the graph area into as many parts necessary based on the data.

As the graph becomes more complex when there are multiple cells within the same page, the graph area might become more complex with the incorporation of gutters, cell headers, and sidebars, as seen in Figure 1-5.

Figure 1-5: Complex Graph Area

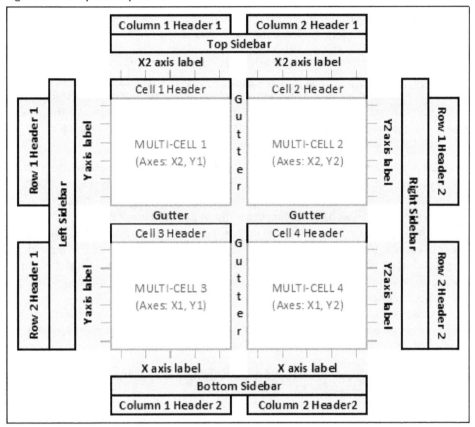

1.1.2.2 Layouts for GTL

There are several types of layouts available in GTL. What needs to be graphed determines the type of layout. These layouts can be classified into one of three categories mentioned previously, which are: single-cell, multi-cell pre-defined, and multi-cell data-driven.

Within the single-cell graph, there are five different layouts. All single-cell layouts, with the exception of OVERLAY3D, are 2-D plots.

- **OVERLAY:** General layout that allows for the plotting of one or more 2-D plots in the same cell provided all plots can share the same axes type.

- **OVERLAYEQUATED:** Similar to OVERLAY with the differences being that the axes can only be linear, the display distance on both axes is the same for equal data, and the aspect ratio of the plot display and plot data are equal.

- **PROTOTYPE:** Layout that creates a prototype of a plot that will be used repeatedly based on the data. Since the number of repeats is based on the data, it can only be used with DATAPANEL or DATALATTICE.

- **REGION:** General plot that does not use axes.

- **OVERLAY3D:** General layout that allows for the plotting of one or more 3-D plots in the same cell provided all plots can share the same axes type.

For pre-defined multi-cell graphs, there are two layouts. Both layouts are 2-D plots.

- **GRIDDED:** A simple multi-cell layout with all cells pre-defined. The grid is determined based on the number of rows and columns that are specified in the options. The cells have the same proportion in regard to height and width and how much space is allocated for the cell content is automatically determined.

- **LATTICE:** A complex multi-cell layout with all cells pre-defined. Similar to GRIDDED, the grid is determined by the number of rows and columns specified in the options. LATTICE is very flexible in that it allows each cell to have different heights and widths. The size of the cells can be specified by indicating the width of each column (COLUMNWEIGHTS) or the height of each row (ROWWEIGHTS) or indicating both width and height. The size of the cells can also be automatically determined based on whether or not the width or height should be equally divided among all columns or among all rows, respectively. In addition, the width or height can be the maximum preferred width from a vertically one-dimensional plot or maximum preferred height from a horizontally one-dimensional plot with the other columns and rows that do not contain the vertically or horizontally one-dimensional plot having an equal width or height from the remaining space.

Both GRIDDED and LATTICE layouts can be used with OVERLAY layout in order to create other types of layouts by nesting within each cell.

Multi-cell data-driven graphs also have two layouts that are both 2-D plots.

- **DATAPANEL:** Uses the PROTOTYPE layout to display a panel of similar graphs that are based on the data. The number of cells produced is determined by the crossings of n classification variables. In order words, each unique value for each classification variable is paired with each unique value of the other classification variable. For example, if you have two classification variables and one variable has three levels and the other variable has four levels, then there would be 12 crossings, that is, each level in the first variable is crossed with each level in the second variable.

- **DATALATTICE:** Similar to DATAPANEL with the difference being that the number of cells is based on crossings of 1 or 2 classification variables.

In addition, there are other types of layouts that do not necessarily pertain to the number of cells.

- **GLOBALLEGEND:** This allows for the creation of one legend using multiple discrete legends or merged legends placed at the bottom of the graph. A continuous legend is not supported within a GLOBALLEGEND. In addition, there is only one GLOBALLEGEND statement allowed per template.

- **INNERMARGIN:** Can only be used inside an OVERLAY or PROTOTYPE layout. It reserves space within those two layouts to provide a nested block plot or axis table. Note that only a BLOCKPLOT or AXISTABLE statement can be used within an INNERMARGIN.

With the exception of OVERLAYEQUATED, OVERLAY3D and REGION, there is at least one example illustrating the different layouts throughout the book.

1.1.2.3 General Syntax for GTL

Before discussing the general syntax for the layouts, you need to understand the syntax for the TEMPLATE procedure. All templates created for the purpose of producing a graph will have the structure illustrated in Program 1-4.

Matange (2013) points out that using GTL to create a graph is a two-step process:

1. First, you need to define the structure of the graph using the STATGRAPH template. In the creation of the template, no graph is actually produced.
2. Second, you need to associate the data in order to render the template that will produce the graph.

Program 1-4 illustrates the basic syntax of the TEMPLATE procedure, the first step in using GTL, with a brief explanation of each component provided.

Program 1-4: TEMPLATE Procedure Syntax

```
proc template;
  define statgraph templatename;    ❶
     begingraph / <options>;    ❷
           layout type / <options>;    ❸

              <... GTL SAS code ...>    ❹

           endlayout;    ❸
     endgraph;    ❷
  end;    ❶
run;
```

❶ The first step to creating a custom graph is by defining a STATGRAPH template within PROC TEMPLATE. This allows for the "opening" of the template that you use to define your graph. Each custom template starts with "define statgraph ***templatename***" that enables you to provide a name for the custom template. The template name is used when rendering the graph. Since you opened the template, you must also close the template, in order to do so you need a corresponding END.

❷   Each STATGRAPH definition has at most one BEGINGRAPH statement, which is the signal that indicates the various components of the custom template are specified within the block. BEGINGRAPH has a corresponding ENDGRAPH statement, which signals the end of the graph template definition.

❸   LAYOUT enables you to specify the type of layout you want and assign the necessary options. Layouts can be nested depending on whether LATTICE, GRIDDED, DATAPANEL or DATALATTICE is specified in the statement. For each LAYOUT, you need to signal the end of the layout with ENDLAYOUT. General syntax on the layouts is explained in Section 1.1.2.4 *General Syntax for Layouts*.

❹   The majority of the work takes place within the LAYOUT block(s). The type of layout(s) and plot(s) drives how this portion is programmed. If there are multiple layouts or nested layouts, then there needs to be a corresponding ENDLAYOUT statement for each. As many plot statements and their corresponding options needed to build the graph should be encompassed within the corresponding LAYOUT.

The template created describes the graph. It is a program that is stored and compiled to be accessed later. As mentioned in "What's Your Favorite Color? Controlling the Appearance of a Graph" (Watson, 2020), the templates that are created by default are stored in SASUSER.TEMPLAT. However, you can make the template available to others by saving it to a permanent location by modifying the ODS PATH statement.

Program 1-5: Saving a Template to a Permanent Location

```
libname ADAM "C:\Users\gonza\Desktop\GTL_Book_Harris_Watson\Data\ADaM";

ods path ADAM.TEMPLAT (update) SASUSER.TEMPLAT (read) SASHELP.TMPLMST (read); ❶
```

❶   In order to save the graph template that you created to a permanent location for others to use, the ODS PATH statement needs to be updated so that it lists the location of where it should reside. SAS searches the template stores in the order in which they are listed. If no modification is made to the ODS PATH, then SAS will search in SASUSER.TEMPLAT first and then SASHELP.TMPLMST. The UPDATE option allows new templates to be written to that template store as well as allowing templates to be read from it.

As stated earlier, creating a graph with GTL is a two-step process because the template itself does not actually produce a graph. Thus, the second step is to render the graph. This is done by associating the data with the template in the SGRENDER procedure. It does not matter what the data source is as long as the data has all the necessary variables that are referenced in the template.

Program 1-6: Associating the Data with the Graph Template

```
proc sgrender data = datasetname ❶   template = templatename;   ❷
    <optional SAS statements>;
run;
```

❶ The name of the data set that contains all the necessary variables that are to be associated with the graph template.

❷ The name of the graph template that will be rendered to produce the actual graph. If the template is saved to a permanent location, then the ODS PATH should have the location specified.

Below are a few examples of other SAS statements that can be used in the SGRENDER procedure.

- BY statement to allow rendering by different groups.

- FORMAT statement to allow data to be formatted without losing the order of the data. This can be beneficial if the data needs to be in a specific order, but If you were to sort by the labels, you would lose the ordering.

- LABEL statement to allow X and Y-axis labels to be defined if they are not defined in the template. This could be beneficial if the template is used for producing multiple outputs but with different X and Y-axis labels.

1.1.2.4 General Syntax for Layouts

PROC TEMPLATE enables you to customize your SAS output. The customization can include things such as color and fonts, which is part of the style. In addition, PROC TEMPLATE is used to specify the necessary GTL statements in order to define the structure of the graph. This includes specifying the layout(s). Each layout is initiated with a LAYOUT statement and terminated with an ENDLAYOUT statement regardless of which layout you use.

The syntax for all the layouts except for the two that are data-driven is similar but will have varying options. The general syntax is simple, as illustrated in Program 1-7.

Program 1-7: General Syntax for Majority of the Layouts

```
layout type </options>;
   <... GTL SAS code ...>
endlayout;
```

The two layouts that are data-driven require that the PROTOTYPE layout be nested within the data-driven layouts as illustrated in Program 1-8 and Program 1-9.

Program 1-8: General Syntax for DATALATTICE

```
layout DATALATTICE rowvar ❶ = class-variable
                    columnvar ❶ = class-variable </option(s)>;
   layout prototype </options>;  ❷
      <... GTL SAS code ...>
   endlayout;
endlayout;
```

❶ Either ROWVAR and/or COLUMNVAR must be specified when using DATALATTICE. Both can also be specified. ROWVAR represents the classification variable for the rows, while COLUMNVAR represents the classification variable for the columns.

❷ The PROTOTYPE layout is used to create the structure of the graph that is used repeatedly based on the data. The general syntax for PROTOTYPE follows the same syntax as shown in Program 1-7.

Program 1-9: General Syntax for DATAPANEL

```
layout DATAPANEL classvars = (class-var1...class-varn)  ❶ </option(s)>;
    layout prototype </option(s)>;
        <... GTL SAS code ...>
    endlayout;
endlayout;
```

❶ CLASSVARS is a list of the classification variables that are to be used to generate the panel. A cell is created for each crossing of the classification variables.

1.1.2.5 Syntax for LATTICE Layout

While the general syntax for LATTICE is similar to the others in its simplest state, as shown in Program 1-7, depending on the desired graph, the syntax used can be more in-depth. LATTICE allows for different axes for each portion of the graph but also allows for the sharing of axes. When using the LATTICE layout, by default, each cell within the lattice has its own internal axes. However, if your cells within the lattice have a common scale, then you can create row and/or column axes that the various cells can share. Program 1-10 illustrates the syntax that would be used when you want to control the axes.

Program 1-10: Controlling Axes Syntax for LATTICE

```
layout LATTICE </options>;
    <... GTL SAS code ...>
    <columnaxes </options>;     ❶
        columnaxis / axis-option(s);
        <... GTL SAS code ...>
    endcolumnaxes;>
    <column2axes </options>;    ❶
        columnaxis / axis-option(s);
        <... GTL SAS code ...>
    endcolumn2axes;>

    <rowaxes </options>;        ❷
        rowaxis / axis-option(s);
        <... GTL SAS code ...>
    endrowaxes;>
    <row2axes </options>;       ❷
        rowaxis / axis-option(s);
        <... GTL SAS code ...>
    endrow2axes;>

    <columnheaders;       ❸
        <... GTL SAS code ...>
    endcolumnheaders;>
    <column2headers;       ❸
        <... GTL SAS code ...>
    endcolumn2headers;>
```

```
    <rowheaders;      ❹
        <... GTL SAS code ...>
    endrowheaders;>
    <row2headers;     ❹
        <... GTL SAS code ...>
    endrow2headers;>

    <sidebar </options>;      ❺
        <... GTL SAS code ...>
    endsidebar;>
endlayout;
```

❶ COLUMNAXES and COLUMN2AXES are used to control the X and X2 axes, respectively. However, if all parts of the graph are to have the same axes, then only one set of axes statements would need to be defined.

❷ ROWAXES and ROW2AXES are used to control the Y and Y2 axes, respectively. Similar to COLUMN(2)AXES, if all parts have the same axes, then only one set needs to be defined.

❸ COLUMNHEADERS displays the header that is to be displayed at the bottom of the graph. This is considered the primary header. In addition to COLUMNHEADERS, you can have COLUMN2HEADERS, which is the secondary header that is displayed at the top.

❹ Like COLUMNHEADERS, ROWHEADERS displays the primary header on the left side of the graph while ROW2HEADERS, which displays the secondary header on the right side.

❺ SIDEBAR is useful for adding information that can span across columns or rows, and up to four sidebars can be specified, one for the bottom, top and each side. For example, you can use the SIDEBAR statement to specify a text that spans across the columns or rows, or it can be used to display legends.

1.2 Reason Why Clinical Graphs Are Needed and What They Should Convey

The majority of the output produced for clinical data is provided in the form of a listing or table. Tables can contain things such as summary statistics over time or confidence intervals as well as *p*-values to test if the difference between two groups of data are significant. While this information is nice in a tabular form, sometimes it does not fully convey the message that you want to convey. A lot of individuals are more apt to understand the meaning of something if it is displayed as a picture. In other words, a lot of people are more visual. They want to see pictures to get a better grasp of what the message is. Because of this, sometimes it is better to display clinical data in a graph.

With a clinical graph, you can better understand mean values over time. In addition, with the clinical graph, you can overlay different types of summary statistics to get a better "picture." For example, you can have a series plot of the mean change from baseline over time with whiskers that show +/- X standard deviations from the mean. This not only shows you how the mean value is changing over time but also shows the spread of each mean value. Granted, all this information can be captured in a table format, but reading each number and flipping through numerous pages makes it easier to miss a key point. However, with the visual representation, the eye is immediately drawn to the part that does not seem to be in alignment with the rest of the data. In addition, with a graphical representation of the data, you can showcase results of different data on one page. For example, as illustrated in Chapter 8, you can overlay medication usage with the change of a particular lab value over time to help monitor if the medication had an effect on the lab result.

Regarding what you should convey in a clinical graph, it depends on the message you want to deliver. Thus, when determining what type of graph you want, you need to ask yourself, "What is the story I am trying to tell?" If you are looking to see the effect that a medication has on particular lab values, then you would need to ensure you have a graph that shows the lab values over time with the medication taken plotted on the same graph. Or, if you are interested in specific adverse events (AEs) and the lab results that are related to those adverse events, then you need to think about the best way to illustrate that. For example, would you display that as a bar chart of the number of AEs at different levels of the mean lab result or would you display that as a box plot of the mean lab values with a scatterplot of AEs overlaid?

1.3 Chapter Summary

While SAS has procedures that produce nice graphs, understanding the basics of ODS Graphics can help you to modify those features that you need to change in order to create the desired output or to help you produce truly custom graphs. You have the ability to modify the template from the procedure-defined templates or create your own graph template. If you know what type of message you want to convey, you can create clinical graphs that will help with understanding your data.

References

Harris, K., & Watson, R. (2018). "Great Time to Learn GTL." Proceedings of the Annual Pharmaceutical SAS Users Group Conference. Seattle: PharmaSUG.

Matange, S. (2013). *Getting Started with the Graph Template Language in SAS®: Examples, Tips, and Techniques for Creating Custom Graphs*. Cary, NC: SAS Institute Inc.

Matange, S. (2016). *Clinical Graphs Using SAS®*. Cary, NC: SAS Institute Inc.

SAS Institute Inc. (2016). *SAS® 9.4 Graph Template Language: User's Guide, Fifth Edition*. Cary, NC: SAS Institute Inc. Retrieved from https://documentation.sas.com/api/docsets/grstatug/9.4/content/grstatug.pdf?locale=en

Watson, R. (2019). "Great Time to Learn GTL: A Step-by-Step Approach to Creating the Impossible." Proceedings of the *SAS Global Forum 2019 Conference* (p. 1). Cary, NC: SAS Institute Inc. Dallas: SAS Institute Inc.

Watson, R. (2020). "What's Your Favorite Color? Controlling the Appearance of a Graph." Proceedings of the *SAS Global Forum 2020 Conference*. Cary, NC: SAS Institute Inc.

Chapter 2: Using SAS to Generate Graphs for Adverse Events

2.1 Introduction to Adverse Events

The International Conference on Harmonization (ICH) has produced a guideline on the structure and content of clinical study reports. This guideline is known as the "ICH E3 Guideline: Structure and Content of Clinical Study Reports." Many pharmaceutical companies adopt this guideline when creating a clinical study report (CSR); therefore, knowing how to create figures that can fulfill some of the recommended outputs is valuable.

CSRs are submitted to the Food and Drug Administration (FDA) in order for the drug, biologic, or medical device to gain approval. If the FDA approves the product, then the product can be marketed within the US. There are many agencies around the world that are responsible for approving and rejecting drug applications within their country. For example, in the UK, the responsible agency is the Medicines and Healthcare Products Regulatory Agency (MHRA), and in Germany, a combination of the Federal Institute for Drugs and Medical Devices and The Ministry of Health agencies are responsible.

Submitting a drug to the German market requires different information compared to submitting a drug to the US market. For example, for the German market, they request time-to-event analysis to be done on the adverse events for oncology treatments.

This chapter demonstrates how you can use SAS to produce visualizations on adverse events (AEs) data regardless of the regulatory agency, in order to help evaluate the safety of a study.

The source data used in these examples are predominantly in Study Data Tabulation Model (SDTM) or Analysis Data Model (ADaM) format. Either fictional data or the SDTM/ADaM Pilot Project, which is available to members on the Clinical Data Interchange Standards Consortium (CDISC) website is the source data for the examples illustrated. SDTM domains are used both explicitly and implicitly in the examples within this chapter. The SDTM domain used in this chapter is the Adverse Events (AE) domain, and the ADaM data set used in the examples is the *Analysis Dataset for Adverse Events* (ADAE) data set.

The data sets from the CDISC SDTM/ADaM Pilot Project were used as the source data for the examples in this chapter. The CDISC SDTM/ADaM Pilot Project was a prospective, randomized, multi-center, double-blind, placebo-controlled, parallel-group study. The objectives of the study were to evaluate the efficacy and safety of transdermal xanomeline, 50 cm2 and 75 cm2, and placebo in patients with mild to moderate Alzheimer's disease.

A total of 254 patients were randomized and entered the double-blind treatment phase. The number of patients randomized to each treatment arm was as follows: 86 to placebo, 84 to the xanomeline low dose treatment group (50 cm2), and 84 to the xanomeline high dose treatment group (75 cm2).

2.2 Reports for Adverse Events

The ICH recommends in section 12.2 of the guidelines that, for the evaluation of adverse events, a display of AEs be made available to regulatory authorities when requested.

The ICH recommends that all AEs occurring after initiation of study treatments be displayed in summary tables. These AEs should be grouped by body system. It might be helpful to also divide the AEs into defined severity categories such as Mild, Moderate and Severe, and/or divide the AEs into those considered treatment-related and those considered not related. In these summary tables, it is useful to identify *treatment emergent adverse events* (TEAEs). Usually (for a simple randomized study), TEAEs are defined as events that occurred after baseline that were not present at baseline, and also events that worsened after baseline. The actual definition of a TEAE is dependent on the study and the client. For example, some clients have a certain cut-off point where an AE that originated after treatment or study discontinuation can be a TEAE if it is prior to the cut-off point, such as AEs that occurred 30 days after the patient has discontinued treatment. Other clients might not have a cut-off, or the cut-off might vary on the number of days after the patient has discontinued treatment.

In regard to the use of the recommended summary tables, as mentioned above, the ICH E3 guideline that is referred to here was finalized in 1995. When the current ICH E3 guideline was first instituted, the ability to create quality graphical displays was limited. However, as technology evolves, so does the ability to produce graphs that nicely complement the summary tables. In some instances, producing a graphical display is more informative than producing a summary table.

This section illustrates how the AE figures can be accomplished with SAS. To create the AE figures, GTL is used in conjunction with the ADAE data set. Because ADaM data set structures are less predominant than SDTM domains, you might not have an ADAE data set to use, although you will likely have data on the AE domain. For illustration, the only difference between the ADaM data set (ADAE) and the SDTM domain (AE), is the treatment emergent flag in ADAE that is used to identify TEAEs. If an ADAE data set is not available, then the AE domain can be used to produce the graphs illustrated in this chapter. If you need to produce the graphs in this chapter using the AE domain, a temporary data set would need to be created first that flagged the TEAEs.

2.3 Adverse Events MedDRA Coding

The ICH recommends grouping the AEs by body system. In the upcoming examples, a body system is also known as *system organ class* (SOC). In the following examples, the AEs were coded according to the *Medical Dictionary for Regulatory Activities* (MedDRA), and the examples in this chapter use MedDRA version 14.0.

In general, MedDRA has become widely recognized as a global standard in the coding of AEs. AEs are recorded in textual or "verbatim" terms. This verbatim term is a short description of the event and is generally written in free text on the case report form (CRF). Verbatim terms are then processed through a coding dictionary so that similar verbatim terms are grouped together by classification of each into a hierarchy of medical granularity. For example, if a verbatim term was recorded as "stomach virus" using MedDRA (version 14.0), then this verbatim term would result in a SOC of "Infections and infestations" and a preferred term (PT) of "Gastroenteritis viral." In the following examples, the MedDRA PT is used interchangeably with the term "adverse event."

In MedDRA version 14.0, there are 26 SOCs. The SOC "Neoplasms Benign, Malignant and Unspecified (Incl Cysts and Polyps)" has the longest name, consisting of 67 characters. The length of the longest SOC name is used to gauge the size the text should be in the plot and/or the size of the plot. This length is used to confirm that the full name can be accommodated on the plot. Because there are 26 SOCs, using a panel consisting of 2 SOCs per output is useful for reducing the number of outputs and will create 13 outputs. In addition, outputting the SOCs in two columns, which are side by side, allows for a comparison of the two SOCs that are on the same output.

2.4 Toxicity Grade for Adverse Events

AEs can have different levels of severity. The AE's severity is usually assessed and recorded by the investigator using the National Cancer Institute (NCI) Common Terminology Criteria for Adverse Events (CTCAE). The NCI-CTCAE provides a grading (severity) scale for each adverse event term.

- Grade 1 is a mild adverse event.
- Grade 2 is moderate.
- Grade 3 is severe.
- Grade 4 is life threatening.
- Grade 5 means that a death is related to the adverse event.

Not all Grades are appropriate for all AEs. Therefore, some AEs are listed with fewer than five options for Grade selection. For example, Grade 5 (Death) is not appropriate for some AEs, and therefore is not an option.

Table 2-1 provides a more detailed description of each of the CTCAE grades. In the table below, instrumental Activities of Daily Life (ADL) refer to preparing meals, shopping for groceries or clothes, using the telephone, and managing money, etc. Self-care ADL refer to bathing, dressing and undressing, feeding self, using the toilet, taking medications, and not being bedridden.

Table 2-1: CTCAE Gradings

Grade	Grade Severity
1	Mild; asymptomatic or mild symptoms; clinical or diagnostic observations only; intervention not indicated.
2	Moderate; minimal, local or non-invasive intervention indicated; limiting age-appropriate instrumental ADL.
3	Severe or medically significant but not immediately life-threatening; hospitalization or prolongation of hospitalization indicated; disabling; limiting self-care ADL.
4	Life-threatening consequences; urgent intervention indicated.
5	Death related to AE.

2.5 Report for Maximum Grade 1 Adverse Events by Treatment Group

This section concentrates on reporting AEs for the xanomeline low-dose treatment group. Once you know how to report AEs for one treatment, then you will know how to report the AEs for all the various treatment groups.

Reporting the AEs for each grade is very useful to identify the severity of the different AEs. Calculating each patient's maximum adverse event grade is useful for comparing AEs within treatments and between treatments. To Report Maximum Grade 1 AEs by Treatment Groups, you need to create a modified data set that has the appropriate calculations. The following steps below are applied to the ADAE data set so that a data set is created with the appropriate calculations. The new data set is named ADAESUMMARY1.

1. The maximum AEs grade each patient had is calculated.
2. The AEs with a maximum grade of 1 are selected.
3. The number of patients in each AEs are calculated.
4. Percentages are calculated by dividing the number of patients in each adverse by the number of patients that took the same treatment.

2.5.1 Illustrating Snippet of Adverse Event Data

Display 2-1 is a snippet of the data, which is used in Program 2-1. The variables TRTAN and TRTA contain the treatment information. AESOC and ORDER contain the SOC information, and the order that the SOCs should be displayed. The ADECOD variable contains the AE preferred term. Since the data used in Program 2-1 contains only results that had a maximum grade of Grade 1, then only data with Grade = "Grade 1" is displayed. The display below is a result of manipulations that were performed on the ADaM ADAE data set, so the columns in the data set below do not represent how the ADaM and/or SDTM data set will exist in a data transfer.

Display 2-1: Data Snippet (ADAESUMMARY1)

TRTAN	TRTA	AESOC	COUNTN	PERCENTAGE	N	ADECOD	YLABEL	ORDER	GRADE
0	Placebo	CARDIAC DISORDERS	9	10.47	86	Total	11.47	5	Grade 1
0	Placebo	CARDIAC DISORDERS	1	1.16	86	Atrioventricular Block First Degree	2.16	5	Grade 1
0	Placebo	CARDIAC DISORDERS	1	1.16	86	Ventricular Hypertrophy	2.16	5	Grade 1
0	Placebo	CARDIAC DISORDERS	1	1.16	86	Atrioventricular Block Second Degree	2.16	5	Grade 1
0	Placebo	CARDIAC DISORDERS	1	1.16	86	Sinus Arrhythmia	2.16	5	Grade 1

2.5.2 Producing Frequency Plot of Adverse Events for Each System Organ Class

Program 2-1 contains code that can produce an AE figure. The program achieves the figure by producing an output of the percentage of patients that had an adverse event for each SOC. Program 2-1 uses the ADAESUMMARY1 data set. Therefore, Program 2-1 creates a frequency plot of maximum grade 1 AEs for each SOC.

Because of the nature of the data referenced in Program 2-1, a template was created with PROC TEMPLATE so that the font sizes are smaller on the outputs, which enables display of the full name of the SOCs. In addition, the smaller font size allows a large number of PTs to be displayed on the X axis. The template that is being created is named AEPanelStyle. The AEPanelStyle references the CustomSapphire style template as the parent template, and the CustomSapphire template is the default template that the outputs in this book use. AEPanelStyle is used to create Output 2-1, Output 2-2, and Output 2-3. The CustomSapphire style can be found at https://github.com/rwatson724/SAS-Graphs-Clinical-Trials-Example.

Program 2-1: AEs by SOC

```
proc template;
    define style styles.AEPanelStyle; ❶
        parent = CustomSapphire;
        style GraphFonts  from GraphFonts /
            'GraphValueFont' = ("<sans-serif>, <MTsans-serif>",5pt)
            'GraphLabelFont' = ("<sans-serif>, <MTsans-serif>",6pt)
            'GraphDataFont' = ("<sans-serif>, <MTsans-serif>",5pt)
            'GraphTitleFont' = ("<sans-serif>, <MTsans-serif>, sans-serif",7pt, bold);
    end;
run;

ods listing gpath = "C:\GTL_Examples\Chapter2\output\AE\ALL\" image_dpi = 300
    style = aepanelstyle;

ods graphics / reset = all imagename = "Output2_1" imagefmt = png height = 3.33in
    width = 5in; ❷

proc template;
    define statgraph AEDatapanelSOC; ❸
        mvar trt2n; ❹
        dynamic _byval_; ❾
        begingraph;
            entrytitle "Actual Treatment = " _byval_ ", N = " trt2n; ❾
            entryfootnote halign=left "Note: The numbers on top of the bars are count of
            patients and not percentages" / textattrs = (size = 6pt);

            layout datapanel classvars = (order) / ❺
                columns = 2 rows = 1  ❺
                columndatarange = union  ❺
                headerlabeldisplay = value  ❺
                headerlabelattrs = (size = 4px) ❺
                columnaxisopts = (label = "Total and MedDRA Preferred Term"
                type = discrete) ❺
                cellwidthmin = 45px cellheightmin = 45px; ❺
```

```
                layout prototype; ❻
                    barchart x = aedecod y = percentage / barwidth = 0.8; ❼
                    scatterplot x = aedecod y = ylabel / markercharacter = countn; ❽
                endlayout;

            endlayout;

        endgraph;
    end;
run;

proc sgrender data = tfldata.adaesummary1 template = AEDatapanelSOC; ❾
    where trtan = 54; ❾
    by trta; ❾
    format order aefmt.; ❾
run;
```

❶ A new style template is defined. The name of the template is AEPanelStyle. The main reason that this template is created is because the AE preferred term names are quite long, and therefore in order for the names to adequately fit on the plot, a smaller text size is defined for the GraphValueFont. In other areas of the graph, it is also beneficial for the text to have a smaller size, and therefore the text attributes for GraphLabelFont, GraphDataFont, and GraphTitleFont are also made smaller.

❷ The IMAGE_DPI option in ODS LISTING specifies the number of dots per inch. The default resolution of graphs created with the LISTING destinations is 96 DPI. An increase in resolution often improves the quality of the graphs, but it also increases the size of the image file. For this graph, the DPI was set to 150 instead of the default 96 to improve the quality of the graph, and to help ensure that the graph is clear when viewed from a distance. For subsequent graphs, the default DPI has also been overridden for the same reasons. The HEIGHT and WIDTH options in ODS GRAPHICS specify that the height and width of the output be 3.33 and 5 inches, respectively. Using the IMAGE_DPI, HEIGHT and WIDTH options as specified, in combination with the AEPanelStyle template, is necessary for the display of the output, particularly when each output displays two SOCs side-by-side.

❸ GTL is used to allow for the creation of a custom template. In this case, GTL is used to create and temporarily store the STATGRAPH template AEDatapanelSOC.

❹ The MVAR statement obtains the macro variable value from the macro variable TRT2N and stores the value as a character value. The macro variable *TRT2N* contains the number of patients randomized to the treatment xanomeline low dose, which is 84. The advantage of using the MVAR statement in the template to define a symbol that references a macro variable rather than using a macro variable, such as *&TRT2N* within the template, are:

- You do not need to embed your PROC TEMPLATE code within a macro for the template to be able to use the macro variables.

- The symbols that reference the macro variables resolve at execution rather than resolve at compilation. Resolving at execution rather than at compilation means that your macro variables can be created after your PROC TEMPLATE code. If you would have used *&TRT2N* in the AEDatapanelSOC template, before you created the TRT2N macro variable, then you would have seen the following warning: "WARNING: Apparent symbolic reference TRT2N not resolved." And if you attempted to create the plot, you would see &TRT2N on your output instead of the value that you wanted to see, that is, 84.

- When you are using the symbols that reference a macro variable to produce titles, you do not need to add quotation marks around the symbol.

❺ The AEDatapanelSOC template consists of two cells, one for each SOC.

- The number of cells needed is defined by the COLUMNS = 2 and ROWS = 1 options in the LAYOUT DATAPANEL statement.

- COLUMNDATARANGE = UNION allows the data on the X axis to be scaled separately for each column. For example, each panel does not have exactly the same X-axis tick values; thus, the scaling of the data range on the X axis does not span across multiple panels.

- The HEADERLABELDISPLAY = value is used to display only the SOC as opposed to also including the variable name.

- Because the SOCs can be quite long the HEADERLABELATTRS is used to restrict the size to 4px.

- CELLWIDTHMIN and CELLWIDTHMAX is used to control the minimum dimensions for each cell. If the cells are too small, then the data and the labels within the header might not be displayed clearly.

❻ LAYOUT PROTOTYPE specifies that each cell has a BARCHART and SCATTERPLOT.

❼ The BARCHART statement plots the percentage of patients that have a maximum grade of 1 for the adverse event.

❽ The SCATTERPLOT statement in conjunction with the MARKERCHARACTER option displays the number of patients that have a maximum grade of 1 for the adverse event on top of the bar chart.

❾ PROC SGRENDER associates the AEDatapanelSOC template with the ADAESUMMARY1 data set, and outputs the graph. Within PROC SGRENDER several other SAS language elements can be used to produce the desired graph.

- The WHERE statement in PROC SGRENDER subsets the data to patients treated with xanomeline low dose.

- The BY statement plots each treatment on a separate page/graph, and the actual value of the treatment is used by the dynamic variable _BYVAL_ to display the actual treatment in the title.

- A format was created based on the data so that the panels are ordered by the SOCs, which have the highest percentage of adverse events. The FORMAT statement associates the order with the SOC.

Output 2-1 displays the number of the patients that had a maximum grade of 1 for the indicated AEs within the specified SOCs that are displayed side by side. General Disorders and Administration Site Conditions is the SOC on the left-hand side of the graph, and Skin and Subcutaneous Tissues Disorders is the SOC on the right-hand side. Usually, AE data is not displayed alphabetically but rather by the most frequently occurring AEs, so AEs that occurred most frequently within the patients are displayed first.

The two SOCs below had the highest number of patients with maximum grade 1 AEs for patients who were treated with low dose xanomeline. The total bar in the General Disorders and Administration Site Conditions SOC shows that almost 30% of the patients suffered from at least one maximum grade 1 adverse event within that SOC.

Output 2-1: Frequency Plot of Maximum Grade 1 AE for each SOC – Low-Dose Xanomeline

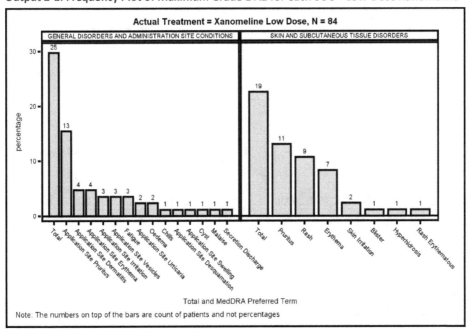

2.5.3 Producing Frequency Plot of Treatment-Related Adverse Events by System Organ Class

The ICH also recognizes that it might be helpful to divide the AEs into those considered treatment-related and those considered not related. Program 2-2 shows how Program 2-1 can be adjusted to produce an adverse event figure describing treatment-related AEs.

Program 2-2: Treatment-Related AEs by SOC

```
proc sgrender data = tfldata.adaesummaryrel1 template = AEDatapanelSOC;  ❶
    where trtan = 54;
    by trta;
    format order aefmtrel.;
run;
```

❶   The data set used in SGRENDER has changed to ADAESUMMARYREL1 to reflect that only the related TEAEs will now be outputted. ADAESUMMARYREL1 is a subset of the ADAESUMMARY1 where AEREL contain the values "POSSIBLE" or "PROBABLE".

Output 2-2 is similar to Output 2-1, but instead displays only the frequency counts of the patients who had a treatment-related maximum grade of 1. Output 2-1 does not take into account the relatedness due to AE and displays both frequency counts of AEs considered treatment-related and AEs considered not treatment-related. As expected, the frequency counts in Output 2-2 are less than or equal to the frequency counts of Output 2-1.

Output 2-2: Frequency Plot of Treatment-Related Maximum Grade 1 AE for each SOC – Low-Dose Xanomeline

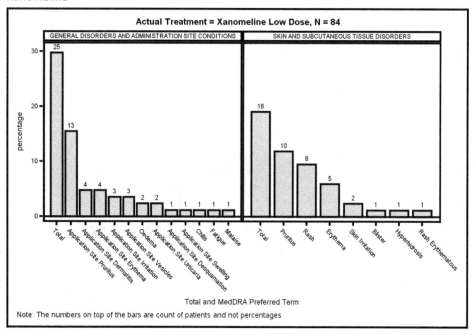

2.5.4 Producing Plot for Both Adverse Events and Treatment-Related Adverse Events by System Organ Class

Sometimes it is necessary to see all possible AEs in relation to the treatment-related AEs while still being able to view two SOCs on the same page. The LAYOUT DATALATTICE can help achieve this because the layout allows for the comparison of two variables. That is, the SOCs and the treatment-relatedness can be compared at the same time.

Program 2-3 shows code that can produce a lattice plot of the AE frequencies. Essentially the output from Program 2-3 is like the combination of Output 2-1 and Output 2-2.

Program 2-3 creates the style template AELatticeStyle. This style has smaller fonts than the AEPanelStyle created in Program 2-1 in order to allow all the required text to be displayed in the lattice row and column headers.

Program 2-3: AEs and Related Adverse Events by SOC

```
proc format;
   value aerel 1 = "All Adverse Events"
               2 = "Treatment Related Adverse Events";
run;

proc template;
   define style styles.AELatticeStyle;
       parent = customsapphire;
       style GraphFonts  from GraphFonts /
           'GraphValueFont' = ("<sans-serif>, <MTsans-serif>",4pt) ❶
           'GraphLabelFont' = ("<sans-serif>, <MTsans-serif>",6pt) ❶
           'GraphDataFont' = ("<sans-serif>, <MTsans-serif>",4pt)  ❶
           'GraphTitleFont' = ("<sans-serif>, <MTsans-serif>,
               sans-serif",7pt, bold); ❶
   end;
run;

proc template;
   define statgraph AEDataLatticeSOC;
   mvar trt2n;
   dynamic _byval_;
   begingraph;
       entrytitle "Actual Treatment = " _byval_ ", N = " trt2n;
       layout datalattice columnvar = aesoc rowvar = trflag /  ❷
           columns = 2 rows = 2  ❸
           headerlabelattrs = GraphValueText(size = 4pt)  ❹
           headerlabeldisplay = value  ❺
           headerlabellocation = inside   ❻
           columnaxisopts = (label = "Total and MedDRA Preferred Term"
               type = discrete)  ❼
           rowaxisopts = (label = "Percentage" tickvalueattrs = GraphValueText); ❽

           layout prototype;
               barchart x = adecod y = percentage / barwidth = 0.8;
               scatterplot x = adecod y = ylabel /  markercharacter = countn;
           endlayout;

       endlayout;
   endgraph;
   end;
run;

proc sgrender data = tfldata.adaesummaryall1 template = AEDataLatticeSOC;
   where trtan = 54 and
         aesoc in ("GENERAL DISORDERS AND ADMINISTRATION SITE CONDITIONS",
                   "IMMUNE SYSTEM DISORDERS");
   by trta;
   format trflag aerel.;   ❾
run;
```

❶ Smaller font sizes were used here, so that the SOC text and the other appropriate labels could be placed in the lattice row and column headers, and still be readable.

❷ The DATALATTICE layout is used when you have one or two classification variables. The two classification variables in this example are AESOC and TRFLAG. The TRFLAG variable is a variable that indicates whether the SOC contains only treatment-related events or if the SOC contains all adverse events. In the ADAESUMMARYALL1 data set, a value of 1 in the TRFLAG variable indicates that the SOC contains all adverse events, and a value of 2 indicates that the SOC contains only treatment-related events. The values are then used with the AEREL format as seen in callout ❾, so that the formatted results can be displayed. The TRFLAG variable helps to ensure that the plots go into the correct cells, so that the top row contains all AEs, and the bottom row contains TEAEs. The data set ADAESUMMARYALL1 is a combination of ADAESUMMARY1 and ADAESUMMARYREL1. The SOC is the column variable, and the treatment-relatedness is the row variable.

❸ The COLUMNS = 2 and ROWS = 2 options, creates a panel with 2 columns and 2 rows.

❹ The HEADERLABELATTRS controls the appearance of the font color and font attributes for the header labels. Using the GRAPHVALUETEXT(size = 4pt) style element option specifies that the header label should use the default text attributes for graph values as specified in the selected style. However, the header label should use the smaller font size of 4pt instead of the default font size specified in the chosen style. Specifying a font size of 4pt makes it easier to fit the entire name of the SOC within the header.

❺ The HEADERLABELDISPLAY has three options to choose from. They are NAMEVALUE, VALUE, and NONE. The NAMEVALUE option uses the format *variable name = value* if there is no variable label, and if there is a variable label, it uses *variable label = value*. The variable AESOC has a label of "Primary System Organ Class", so using the NAMEVALUE option results in a header label of Primary System Organ Class = GENERAL DISORDERS AND ADMINISTRATION SITE CONDITIONS for the GENERAL DISORDERS AND ADMINISTRATION SITE CONDITIONS SOC. Generally, displaying the label is not necessary because it is the same for every cell, and it usually takes up a lot of space. In that case, you can use the VALUE option, to simply display only the value. If you do not want to display anything, then you can use the NONE option and this suppresses all the cell headers.

❻ The HEADERLABELLOCATION = INSIDE option displays the cell header within each cell. The alternative is to use HEADERLABELLOCATION = OUTSIDE to display the row and column headers external to the lattice. The INSIDE option was used because it allowed the SOC headers to have a little more space and display the entire SOC names without truncation.

❼ The COLUMNAXISOPTS options controls the X axis for all columns.

❽ In Outputs 2-1 and 2-2, the Y axis has the label of "percentage". The ROWAXISOPTS specifies that the label for Output 2-3 should be in proper case, so LABEL = "Percentage" is used.

❾ The format statement in PROC SGRENDER associates the TRFLAG with the AEREL format. Therefore, instead of displaying the values 1 and 2, the values "All Adverse Events" and "Treatment Related Adverse Events" are displayed.

Output 2-3 is essentially the combination of Output 2-1 and Output 2-2. A benefit of using Output 2-3 to assess the AEs in each SOC is that doing so enables easy comparison of AEs with treatment-related AEs. Output 2-3 is ordered alphabetically by SOC for simplicity and because this output is mainly to compare within the SOC. That is, all AEs against treatment-related AEs within an SOC.

Output 2-3: Frequency Plot of Maximum Grade 1 AE for each SOC – Xanomeline Low Dose

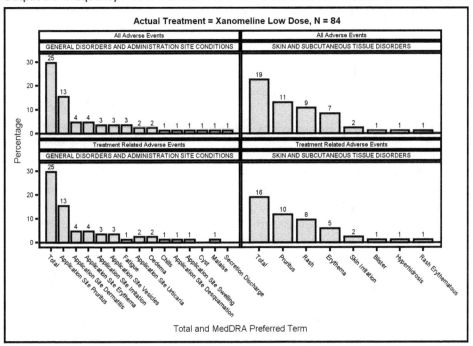

2.5.5 Producing Frequency Plot of Adverse Events Grouped by Maximum Grade for Each System Organ Class

So far, all outputs have been only displaying AEs that had a maximum grade of 1. It is also useful to display the distribution of (maximum) grades associated with each AE. Therefore, it might be necessary to display each grade within an AE as stacked segments within a bar chart. Program 2-4 shows how this can be achieved and contains code that creates a frequency plot of AE grouped by maximum grade for each SOC. Program 2-4 uses a data set named ADAESUMMARYALLSTACKED1. This data set is similar to the data set ADAESUMMARY1; however, this data set contains the maximum grade associated with each AE instead of just the AEs, which had a maximum grade of 1.

As previously mentioned, Program 2-4 will display the distribution of grades associated with each AE. Therefore, it is essential to be able to identify the grades, and this can be done by coloring each grade differently.

Per "What's Your Favorite Color? Controlling the Appearance of a Graph" (Watson, 2020), there are several ways to control the color of each grade and one option is using an attribute map. An attribute map enables you to control not only the color but various other attributes as well, such as line patterns, markers and size for each formatted data value that is specified. When defining your attribute map, if the data is formatted, then the formatted value must be used in the map definition.

One way to create an attribute map is within the template using DISCRETEATTRMAP. If using DISCRETEATTRMAP to define the attribute map, it **must** be placed after the BEGINGRAPH statement but prior to the first layout statement within the template. There are three components to creating an attribute map:

1. The attribute map needs to be initialized. At this step, the attribute map is provided a name. In addition, several options can be specified, such as IGNORECASE, TRIMLEADING, and DISCRETELEGENDENTRYPOLICY. IGNORECASE indicates that when the values are compared that the casing should be ignored. TRIMLEADING is used to remove leading blanks from both the attribute map values and the input data values. DISCRETELEGENDENTRYPOLICY determines whether items in a discrete legend are items in the data or only in the attribute map.

2. After the attribute map is initialized, then the values associated with the attribute map need to be defined. The VALUE statement enables you to specify various attributes such as color, marker, size, and line pattern. Note that if multiple formatted data values share the same attributes, then it is not necessary to specify a VALUE statement for each formatted value. One VALUE statement can be specified listing all the formatted values with a common set of attributes. When listing all formatted values on one VALUE statement, each individual value is to be enclosed in its own set of quotation marks and separated by a blank space. Furthermore, a VALUE statement for OTHER, not enclosed in quotation marks, can be defined to create a category for all other values that are not specified in one of the other VALUE statements. If OTHER is not specified, then any value not defined in the attribute map will be treated as if there was no attribute map defined.

3. The final component is to signal the end of the attribute map definition. This is done with ENDDISCRETEATTRMAP.

Note that multiple attribute maps can be defined within a template. However, they all must be done after the BEGINGRAPH statement and prior to the first layout statement.

Program 2-4: AE Grouped by Maximum Grade

```
proc template;
   define style styles.aepanelstyle;
      parent = customSapphire;
      style GraphFonts  from GraphFonts /
          'GraphValueFont' = ("<sans-serif>, <MTsans-serif>",5pt)
          'GraphLabelFont' = ("<sans-serif>, <MTsans-serif>",7pt)
          'GraphDataFont' = ("<sans-serif>, <MTsans-serif>",5pt)
          'GraphTitleFont' = ("<sans-serif>, <MTsans-serif>,
          sans-serif",7pt, bold);
   end;
run;

proc template;
   define statgraph AEDataPanelGroupSOC;
   mvar trt2n;
   dynamic _byval_;
   begingraph;

   discreteattrmap name="Grade_Color";  ❶
      value "Grade 1" /fillattrs = (color = LightGreen) ❶
         lineattrs = (color = black pattern = solid);
      value "Grade 2" /fillattrs =   (color = DarkGreen) ❶
         lineattrs = (color = black pattern = solid);
      value "Grade 3" /fillattrs =(color = Yellow)  ❶
         lineattrs = (color = black pattern = solid);
      value "Grade 4" /fillattrs = (color = Orange) ❶
         lineattrs = (color = black pattern = solid);
      value "Grade 5" /fillattrs = (color = Red) ❶
         lineattrs = (color = black pattern = solid);
   enddiscreteattrmap;

   discreteattrvar attrvar = grade_map var = grade attrmap = "Grade_Color";  ❷

   entrytitle "Actual Treatment = " _byval_ ", N = " trt2n;
      layout datapanel classvars = (order) / columns = 1 rows = 1
         rowaxisopts = (offsetmin = 0)
         columndatarange = union headerlabelattrs = GraphValueText
         headerlabeldisplay = value
         columnaxisopts = (label = "MedDRA Preferred Term"  type = discrete)
         rowaxisopts = (label = "Percentage")
         cellwidthmin = 60px cellheightmin = 60px;

         layout prototype;
            barchart x = adecod y = percentage / barwidth = 0.8
               group = grade_map grouporder = descending name = "aelegend"; ❸
         endlayout;

         sidebar / align = bottom;
            discretelegend "Grade_Color" / title = "CTCAE Grade:" type = fill
                       displayclipped = true;  ❹
         endsidebar;

      endlayout;
   endgraph;
   end;
run;
```

```
proc sgrender data = adaesummaryallstacked1 template = AEDataPanelGroupSOC;
  where trtan = 54 and
        aesoc = "GENERAL DISORDERS AND ADMINISTRATION SITE CONDITIONS";
  by trta;
  format order aefmt.;
run;
```

❶ DISCRETEATTRMAP and the VALUE statements are used to define the attributes for each Grade. For example, the attributes for "Grade 1" AEs are defined as having a light-green fill color and a black line color, and "Grade 2" AEs are defined as having a dark-green fill color and a black line color. A VALUE statement is needed for each unique value so that a specific color scheme can be used for that value. The specified attributes mean that for a (stacked) bar chart as shown in Output 2-4, "Grade 1" AEs will have a black outline color, and the fill color of the bar will be light green.

❷ DISCRETEATTRVAR is used to associate the attributes that are defined for each value in the DISCRETEATTRMAP section with the variable in the data set, which contains those values. In the ADAESUMMARYALLSTACKED1 data set, the GRADE variable contains the "Grade 1", "Grade 2", and "Grade 3" values, etc. DISCRETEATTRVAR creates an association between the attribute map GRADE_COLOR and the GRADE variable from the data set. This association is called GRADE_MAP. Therefore, when GRADE_MAP is referenced in a plot or legend statement, the correct color scheme is used for the different values of GRADE.

❸ In the group option of the BARCHART plot statement, GRADE_MAP is used to color the different grades in the bar chart as specified in the DISCRETEATTRMAP section.

❹ If a portion of a legend cannot be produced or is truncated (clipped), then by default the legend might be removed from the graph (DISPLAYCLIPPED = FALSE). In order for the legend to still display regardless if part of the legend is "clipped", DISPLAYCLIPPED should be set to TRUE. However, if the font size is adjusted as it is done for the custom style, aepanelstyle, then DISPLAYCLIPPED = TRUE might not be necessary.

Output 2-4 shows the frequency plot of AE grouped by Maximum Grade for the General Disorders and Administration Site Conditions SOC.

Output 2-4: Frequency Plot of AE Grouped by Maximum Grade for Each SOC

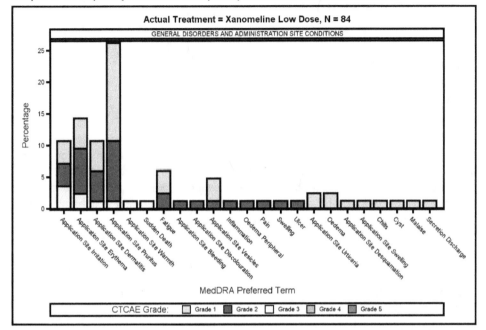

2.6 Report for Time to Adverse Events by Treatment Group

Time-to-event analysis is very common when analyzing overall survival and progression free survival. However, time-to-event analysis can also be done on Adverse Events and provided in submissions. Although time-to-event analysis is illustrated in this chapter, it will be explained in more detail in *Chapter 5: Using SAS to Generate Graphs for Time-to-Event Endpoints*. The data set should follow the CDISC ADaM Basic Data Structure (BDS) for Time-to-Event Analyses. Typically, the parameter used is "Time to First XXXXXX," where XXXXXX could be a number of different items, such as:

- TEAE
- Serious TEAE
- Grade 3 TEAE or higher
- TEAE that led to discontinuation
- Treatment Emergent Headache

In this section, the time-to-first dermatological event is analyzed using Kaplan Meier Curves.

2.6.1 Illustrating Snippet of Time to Event Analysis Data Set

Display 2-2 shows the first 5 rows within the time-to-event analysis data set (ADTTE) data, and the ADTTE data set is used in Program 2-5. For those unfamiliar with CDISC ADaM standards:

- USUBJID contains the Unique Subject Identifier, which is a variable that uniquely identifies a patient within a study and might also uniquely identify a patient across multiple studies if applicable.

- TRTAN and TRTA contain the actual treatment information. The ADTTE data set also contains the TRTPN and TRTP variables, which represent the planned treatment information. Typically, for safety analysis, the actual treatment information is used. The reason being is that the ADaM Implementation Guide (ADaMIG) indicates that a treatment variable is present in the data set. Although, it does not indicate what variable, so if the planned is the same as the actual treatment or if it is an open-label study, then there is no need for actual treatment variables. Currently, TRTAN contains the values 0, 54, and 81. Program 2-5 changes these to 1, 2, and 3 respectively, so that the numbers are sequential, and therefore the numbers will have a consistent order when presented on the graphs.

- PARAM fully describes the value used in the analysis, which in this case is Time to First Dermatological Event in **days**.

- AVAL contains the number of days until the patient had their first dermatological event, or in cases where they did not have a dermatological event, until the patient was censored. The number of days is based on the time from the treatment start date (that is, date first at risk) to date of event. If the event occurred on the treatment start date, then that event would have an AVAL equal to 1. For this example, patients that did not have a dermatological event are censored to their study completion date.

- ADT contains the date of the first event, or the censoring date.

- CNSR identifies whether a patient had an event or was censored. A value of 0 signifies that the patient had the event, and a value of 1 signifies that the patient's AVAL was censored.

- EVENTDESC description offers a description on the event or censored results.

Display 2-2: Data Snippet of ADTTE

USUBJID	TRTA	TRTAN	PARAM	AVAL	ADT	CNSR	EVNTDESC*
01-701-1015	Placebo	0	Time to First Dermatologic Event	2	03-Jan-14	0	Dematologic Event Occurred
01-701-1023	Placebo	0	Time to First Dermatologic Event	3	07-Aug-12	0	Dematologic Event Occured
01-701-1028	Xanomeline High Dose	81	Time to First Dermatologic Event	3	21-Jul-13	0	Dematologic Event Occured
01-701-1033	Xanomeline Low Dose	54	Time to First Dermatologic Event	28	14-Apr-14	1	Study Completion Date
01-701-1034	Xanomeline High Dose	81	Time to First Dermatologic Event	58	27-Aug-14	0	Dematologic Event Occured

*The spellings in this data snippet are based on the values in the CDISC Pilot Study, and no changes were made to the data set.

Program 2-5 shows how you can create a Kaplan Meier (KM) plot to assess the Time-to First Dermatological Event.

Program 2-5: Using KM to Assess the Time to First Dermatological Event

```
data adtte;
   set adam.adtte;
   if trta = "Placebo" then trtan = 1;
   else if trta = "Xanomeline Low Dose" then trtan = 2;
   else if trta = "Xanomeline High Dose" then trtan = 3;
run; ❶

proc lifetest data = adtte plots = survival(atrisk = 0 to 210 by 30);
   time aval * cnsr(1);
   strata trtan;
run; ❷
```

❶ TRTAN value is reassigned so that it is sequential.

❷ The PLOTS option in the LIFETEST Procedure produces the KM plot in Output 2-5. While you might not be specifying any GTL statements, PROC LIFETEST generates the necessary template to produce the graph based on the options. "Procedures such as the Statistical Graphics (SG) procedures dynamically generate GTL templates based on the plot requests made through the procedure syntax" (Watson, 2019). For the purpose of this book we will refer to these types of templates as procedure-driven templates.

Output 2-5 shows the KM curves of the time to first dermatological event, stratified by the treatment. The treatment legend is as follows:

- The placebo group is denoted by the solid blue line, and the number 1.
- The xanomeline low dose group is denoted by the red dashed line, and the number 2.
- The xanomeline high dose group is denoted by the green, more frequent dashed line, and the number 3.

The output shows that the xanomeline low dose and xanomeline high dose had a lot more events than the placebo group because the KM curves are lower for the xanomeline low dose and xanomeline high dose groups. Directly below the KM curves, the graph has a table that shows the number of patients at risk at 30-day intervals, which is produced based on the ATRISK option specified in PROC LIFESTEST.

Output 2-5: KM Plot of Time to First Dermatological Event (in Days) by Treatment

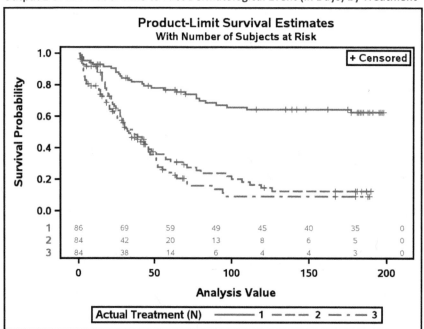

2.6.2 Controlling the PROC LIFETEST Kaplan Meier Plot

You have seen that it is possible to create a KM plot just by using PROC LIFETEST and the PLOTS option. You might now be wondering how you can possibly modify the graph so that you can change the titles, variable labels, or change the tick values on the axes.

If you recall, Chapter 1 explains that one of the reasons to learn GTL (according to Sanjay Matange) was because GTL is the language used behind the SAS analytic procedures for the creation of automatic graphs. Therefore, in order to customize one of those graphs, you will need to understand GTL. In this case, the SAS analytic procedure is PROC LIFETEST, and the automatic graph is the KM plot.

The paper (Kuhfeld and So, "Creating and Customizing the Kaplan-Meier Survival Plot in PROC LIFETEST," 2013) demonstrates at least two ways to modify the default KM plot. In this section, only two methods from the paper are mentioned; however, the primary focus is on the second method. The methods are:

- Controlling the KM by specifying procedure options;
- Controlling the KM by modifying graph templates.

2.6.2.1 Controlling the Kaplan Meier Plot by Specifying Procedure Options

Program 2-5 illustrated how you could control the survival plot through the use of different options. In Program 2-5, the ATRISK option allowed for the production of the at-risk table within the KM plot. If the ATRISK option is removed, then the at-risk table would also be removed. Other procedure options that are available enable you to:

- Change the tick values on the axis;
- Produce confidence bands, such as Hall-Wellner confidence bands;
- Specify the location of the at-risk table;
- Display the *p*-value of a homogeneity test, such as the log-rank *p*-value;
- Suppress the censored values.

Please see the paper (Kuhfeld and So, "Creating and Customizing the Kaplan-Meier Survival Plot in PROC LIFETEST," 2013) for more details about how to control the KM plot by using the procedure options.

2.6.2.2 Controlling the Kaplan Meier Plot by Modifying Graph Templates Using Macros

In Kuhfeld and So's paper, it demonstrates how you can modify the template behind PROC LIFETEST in order to produce the desired KM plot. Modifying the template behind PROC LIFETEST can be achieved by using macros developed by Warren Kuhfeld, which are available in the SAS/STAT documentation available on the SAS Support website (SAS Institute Inc., 2013). The names of the macros are *%SurvivalTemplateRestore* and *%SurvivalTemplate* and they can be used by editing the appropriate macro variables, as Program 2-6 demonstrates. The *%SurvivalTemplateRestore* and *%SurvivalTemplate* macros, are relatively old macros and are used for illustration purposes, and in case there is any legacy code. Newer macros are available, such as the *%ProvideSurvivalMacros* and *%CompileSurvivalTemplates* macros, which is shown in this chapter also.

Program 2-6 shows how you can change the Y-axis label from the default "Survival Probability" label to "Probability." For this example, using "Probability" for the Y-axis label might be a better description of the KM plot, because the output is based on dermatological adverse events, and therefore having a dermatological AE does not impact the patient's survival. Be careful not to confuse this to mean the probability of having an AE. In addition, Program 2-6 shows you how you can modify the KM plot titles.

Program 2-6: Modifying KM Template

```
%SurvivalTemplateRestore; ❶
%let titletext2 = "KM plot of Time-to First Dermatological Event (in Days) by
Treatment"; ❷
%let yoptions = label = "Probability"; ❸
%SurvivalTemplate; ❶/* Compile the templates with */

ods trace on; ❻
proc lifetest data = adtte plots = survival(atrisk = 0 to 210 by 30); ❹
    time aval * cnsr(1);
    strata trtan;
run;
ods trace off; ❻

proc template;
    delete Stat.Lifetest.Graphics.ProductLimitSurvival / store = sasuser.templat;
❺
run;
```

❶ The **%SurvivalTemplateRestore** and **%SurvivalTemplate** macros can be found in the
    SAS/STAT documentation available on the SAS Support website (SAS Institute Inc., 2013).
    The code can be downloaded from the SAS Support website and then run in SAS. The total
    list of macro variables that are available to have their values edited can be seen at the
    beginning of the **%SurvivalTemplateRestore** macro. Page 11 of Kuhfeld and So's "Creating
    and Customizing the Kaplan-Meier Plot in PROC LIFETEST" describes in more detail the
    macro variables, which are available (Kuhfeld and So, 2013). In Program 2-6, the macro
    variables that are edited are *TITLETEXT2* and *YOPTIONS*. If you do not edit any of the macro
    variable values within the **%SurvivalTemplateRestore** macro, then the template behind
    PROC LIFETEST will be very similar to the original PROC LIFETEST template. That is, the KM
    plot will have the same titles and axis labels as it would have had using the original PROC
    LIFETEST template. To conclude, the **%SurvivalTemplateRestore** macro specifies the macro
    variable values for each macro variable. The **%SurvivalTemplate** macro picks up the macro
    variables values which the user specifies, and therefore this is how the KM plot is modified.
    You do not need to edit any of the code within the **%SurvivalTemplate** macro; all you need
    to do is run the macro.

❷ The *TITLETEXT2* macro variable controls the first title when using the STRATA statement in
    PROC LIFETEST. Thus, this macro variable is used to change the first title.

❸ The *YOPTIONS* macro variable is used to change the Y-axis label from the default "Survival
    Probability" to "Probability".

❹ The same PROC LIFETEST statement that was used in Program 2-5 is also used in Program 2-
    6.

❺ Executing the **%SurvivalTemplateRestore** and **%SurvivalTemplate** macros results in the
    compilation of a new template called Stat.Lifetest.Graphics.ProductLimitSurvival and which
    is stored in the Sasuser.Templat item store. This new template is what is now associated
    with PROC LIFETEST, any subsequent KM curves that are produced are now based on this
    new template. In order to remove this new template and go back to the original template
    behind PROC LIFETEST, you can use the DELETE statement in PROC TEMPLATE.

❻ ODS TRACE ON can be used to find out the templates, which are used behind the analytical procedures. To find out the templates that are used in PROC LIFETEST, place ODS TRACE ON right before PROC LIFETEST, and then run the ODS TRACE ON statement and the procedure. In the log, you will see the name, label, template, and path for each output object. For the KM plot, the interest is in the SurvivalPlot output object, and the path for this object is Stat.Lifetest.Graphics.ProductLimitSurvival. The ODS TRACE OFF statement is used to suppress the writing of each output object to the SAS log.

Output 2-6 shows the same KM plot as in Output 2-5. Although in Output 2-6, the first title and the Y-axis label has been updated.

Output 2-6: Producing KM Plot with Modified Template

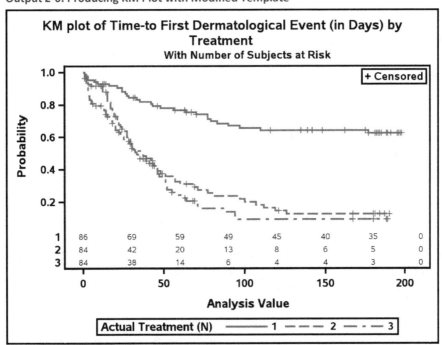

There are other macros that can be used to modify the PROC LIFETEST graph templates. The paper (Kuhfeld and Ying, "Creating and Customizing the Kaplan-Meier Survival Plot in PROC LIFETEST in the SAS/STAT® 13.1 Release," 2014) shows that the macros *%ProvideSurvivalMacros*, and *%CompileSurvivalTemplates* along with defining the appropriate macro variables can be used to modify the KM template. The macros *%ProvideSurvivalMacros* and *%CompileSurvivalTemplates* work in a similar fashion to the *%SurvivalTemplateRestore* and *%SurvivalTemplate* macros. The main difference is that the *%ProvideSurvivalMacros* and *%CompileSurvivalTemplates* macros are newer. For example, with the newer macros, you are able to color your patient's at-risk numbers by a classification group, such as a treatment group. However, with using the older macros, it is difficult to display different patient at-risk colors for the treatment group because the BLOCKPLOT statement is used to create the at-risk tables in the older macros. Nevertheless, these four macros are all used as a way to create a "temporary"

template, which can be modified more easily to produce the desired KM plot when using PROC LIFETEST.

The macros **%ProvideSurvivalMacros** and **%CompileSurvivalTemplates** can be accessed by running the code, which is located in the SAS/STAT documentation available on the SAS Support website (SAS Institute Inc., 2019). An advantage of using the **%ProvideSurvivalMacros** and **%CompileSurvivalTemplates** is that you can easily produce a summary table, which shows the number of events each treatment had and the median time to the first dermatological event. Program 2-7 shows how you can produce the summary table.

Program 2-7: Modifying KM Template in Order to Produce a Summary Table

```
%ProvideSurvivalMacros;  ❶
%let TitleText2 = "KM plot of Time-to First Dermatological Event (in Days) by
Treatment";
%let yOptions = label = "Probability";
%SurvivalSummaryTable;  ❷

%CompileSurvivalTemplates;  ❶

proc lifetest data = adtte plots = survival(atrisk = 0 to 210 by 30);  ❸
    time aval * cnsr(1);
    strata trtan;
run;

proc template;  ❸
    delete Stat.Lifetest.Graphics.ProductLimitSurvival / store = sasuser.templat;
run;
```

❶ The **%ProvideSurvivalMacros** and **%CompileSurvivalTemplates** macros were used in the same way as the **%SurvivalTemplateRestore** and **%SurvivalTemplate** macros. The **%ProvideSurvivalMacros** macro resets the macro variables back to the default macro variable values as specified by the macro. The **%ProvideSurvivalMacros** macro also shows the macro variables, which are available to allow for editing of their values. The **%CompileSurvivalTemplates** macro picks up the macro variables values which the user specifies, and therefore produces the modified KM plot.

❷ The **%SurvivalSummaryTable** macro is used to plot the summary table under the KM graph. Pages 26 and 27 on this paper (Kuhfeld and Ying, 2014) show in more detail how to use the **%SurvivalSummaryTable** and other macro variables.

❸ PROC LIFETEST was used to create the KM plot, and PROC TEMPLATE was used to delete the compiled Stat.Lifetest.Graphics.ProductLimitSurvival template. This code is exactly the same code as in Program 2-6.

Output 2-7 shows the KM Plot with a summary table below the graph. The summary table below the graph shows the number of patients each treatment had, the number of events, the number of censored values, and the median time-to-event in and associated 95% confidence intervals.

Output 2-7: KM Plot with Summary Statistics Table

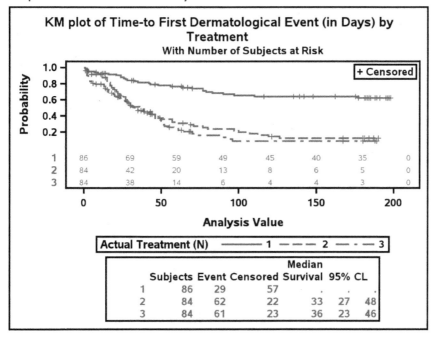

2.6.2.3 Controlling the Kaplan Meier Plot by Directly Modifying Graph Templates

As you have seen, it is possible to use macros to easily edit the KM templates just by editing the macro variable values that are available to **%SurvivalTemplateRestore** and **%ProvideSurvivalMacros**. For the majority of the time, using that approach is sufficient. There might be times when you are not able to produce the desired graph just by editing the macro variable values within those macros; therefore, you might need to edit the template directly by modifying the code within PROC TEMPLATE, which defines Stat.Lifetest.Graphics.ProductLimitSurvival. For example, if you wanted to create a KM plot that has smaller text in the summary table and displays "N" instead of "Subjects", then you can do this by editing the template. The default template can be accessed by using the statement SOURCE Stat.Lifetest.Graphics.ProductLimitSurvival within PROC TEMPLATE. Although, if you recall, a version of the template can be accessed from the SAS/STAT documentation available on the SAS Support website (SAS Institute Inc., 2019). Thus, all that is needed is for the new template that will be associated with PROC LIFETEST, is to be downloaded, modified appropriately to meet the desired needs, and then compiled. This is because within the **%ProvideSurvivalMacros** macro, there are several macros that are used. One of them is the **%CompileSurvivalTemplates** macro and there are several smaller macros within that macro. Making modifications to these smaller macros and re-compiling can help to achieve the desired output. Program 2-8 shows snippets of the code within the PROC TEMPLATE and the **%CompileSurvivalTemplates** macro. Program 2-8 also shows several template modifications that can be done, such as making the text smaller in the summary table and displaying the value "N" instead of "Subjects" within the summary table header. Understanding GTL enables you to know where to make the modifications.

Program 2-8: Directly Modifying KM Template in Order to Produce a Summary Table

```
%macro CompileSurvivalTemplates;
    <... SAS Code ...>

    if (NSTRATA = 1)
        <... SAS Code ...>

    else
        %if &ntitles %then %do; entrytitle &titletext2  / textattrs=(size=10);  ❶
%end;
        %if &ntitles gt 1 %then %do;
            if (EXISTS(SECONDTITLE))
                entrytitle SECONDTITLE / textattrs=GRAPHVALUETEXT;
            endif;
        %end;

            %AtRiskLatticeStart
            layout overlay / xaxisopts=(&xoptions label="Time from First Dose")  ❷
                        yaxisopts=(&yoptions);
                %StmtsTop
                %MultipleStrata
                %StmtsBottom
            endlayout;
            %AtRiskLatticeEnd(class)
        endif;
        <... SAS Code ...>

    %macro SurvTabHeader(multiple);
        %if &multiple %then %do; entry ""; %end;
        entry "";
        entry "";
        entry "";
        entry &r "Median";
        entry "";
        entry "";

        %if &multiple %then %do; entry ""; %end;
        entry &r "N";   ❸
        entry &r "Event";
        entry &r "Censored";
        entry &r "Survival";
        entry &r PctMedianConfid;
        entry halign=left   "CL";
    %mend;

    %macro SurvivalTable;
        <... SAS Code ...>

        if (NSTRATA = 1)
            <... SAS Code ...>

        else
            %SurvTabHeader(1)
            %do i = 1 %to 10;
                %let t = / textattrs=GraphData&i (size=7);   ❹
                dynamic StrVal&i NObs&i NEvent&i Median&i
                        LowerMedian&i UpperMedian&i;
                if (&i <= nstrata)
```

```
            entry &r StrVal&i &t;
            entry &r NObs&i &t;
            entry &r NEvent&i &t;
            entry &r eval(NObs&i-NEvent&i) &t;
            entry &r eval(put(Median&i,&fmt)) &t;
            entry &r eval(put(LowerMedian&i,&fmt)) &t;
            entry &r eval(put(UpperMedian&i,&fmt)) &t;
         endif;
      %end;
   <... SAS Code ...>
   %mend SurvivalTable;
%mend CompileSurvivalTemplates;

%ProvideSurvivalMacros;
%let TitleText2 ="KM plot of Time-to First Dermatological Event (in Days) by
   Treatment";
%let yOptions = label="Probability";
%SurvivalSummaryTable;

%CompileSurvivalTemplates;

proc lifetest data = adtte plots = survival(atrisk = 0 to 210 by 30);
   time aval * cnsr(1);
   strata trtan;
run;

proc template;
   delete Stat.Lifetest.Graphics.ProductLimitSurvival / store=sasuser.templat;
run;
```

❶ Within the *%CompileSurvivalTemplates* macro, you can adjust the font size of the title by specifying the size in the TEXTATTRS option. By specifying TEXTATTRS=(SIZE=10) allows the first title to have a smaller font size.

❷ Within the *%CompileSurvivalTemplates* macro, there are two places where you can specify the X-axis label. This is based on whether you have used the STRATA statement in PROC LIFETEST, and if the stratified variable has at least two values. If your stratified variable has at least two values, then you will need to modify the XAXISOPTS option within the second *%AtRiskLatticeStart* macro in order to update the X-axis label on your KM plot. You can modify the X-axis label by specifying the option LABEL= "Time from First Dose".

❸ Specifying ENTRY &R "N" within the *%SurvTabHeader* macro displays the value "N" instead of "Subjects" in the summary table.

❹ Specifying TEXTATTRS=GRAPHDATA&I(SIZE=7) within the *%SurvivalTable* macro makes the values in the summary table have a smaller font size.

Output 2-8 shows the KM Plot with a smaller title, a different label on the X axis, and different values and font sizes on the summary table, compared to Output 2-7.

Output 2-8: KM Plot with Modified Summary Statistics Table

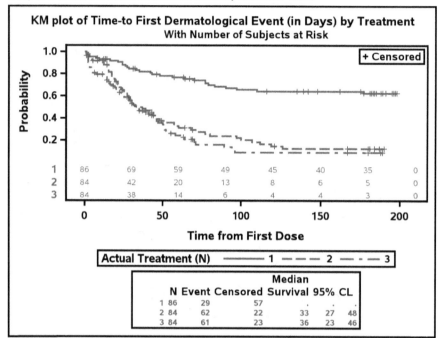

2.7 Treatment Risk Difference for Adverse Events

In Section 2.5, *Reporting Maximum Grade 1 Adverse Events by Treatment Group*, bar charts were shown only for the xanomeline low dose treatment group. It is useful to see bar charts of other treatment groups too, and it is also useful to compare the bar charts between the treatments. One way to compare the differences between the treatments is to plot risk differences. In this context, the risk difference is the difference between the proportions of patients with TEAEs in two of the treatment groups. Display 2-3 shows the adverse event percentage rate of the patients treated with xanomeline low dose and the placebo. The percentage of patients across all TEAEs are also shown, and the risk difference between xanomeline low dose and placebo is also shown.

Display 2-3: Data Snippet of Risk Difference Data (RELDIFFDATA)*

Preferred Term and Total	Order	Xanomeline Low Dose %	Placebo %	Risk Difference
APPLICATION SITE PRURITUS	1	26.2	7.0	19.2
PRURITUS	1	25.0	9.3	15.7
ERYTHEMA	1	16.7	9.3	7.4
RASH	1	15.5	5.8	9.7
APPLICATION SITE ERYTHEMA	1	14.3	3.5	10.8
APPLICATION SITE DERMATITIS	1	10.7	5.8	4.9
APPLICATION SITE IRRITATION	1	10.7	3.5	7.2
TOTAL	2	73.8	33.7	40.1

*Display 2-3 is a data snippet of RELLDIFFDATA. The preferred terms were sorted by the xanomeline low dose percentage difference from largest to smallest, and only the preferred terms, which had a percentage of at least 10%, were kept. Display 2-3 also contains the total row from the RELLDIFFDATA data set.

Program 2-9 shows the code that can be used to plot the risk difference. It is important to appropriately sort the data set before you use the BARCHARTPARM procedure with the HORIZONTAL option. In this instance, appropriate means, sorting the data by descending "Order" and ascending "Risk Difference." The data set needs to be sorted in this manner because using the HORIZONTAL option with the BARCHARTPARM statement plots the data in the reverse order. The values that have a larger risk difference and smaller order number are plotted on the top of the graph.

Program 2-9: Risk Difference

```
proc template;
   define statgraph reldiff;
      begingraph;
         layout overlay / xaxisopts=(label = "Percent Difference (Xanomeline low
                                    dose - Placebo)")
                         yaxisopts=(tickvalueattrs=(size=4.5));
            barchartparm category = aedecod response = xanlow_minus_placebo /
               orient=horizontal; ❶
         endlayout;
      endgraph;
   end;
run;

proc sgrender data = reldiffdata template=reldiff;
run;
```

❶ BARCHARTPARM is used to plot the bar chart of the relative differences. The AEDECOD variable is the "Preferred Term and Total" variable in Display 2-3. The XANLOW_MINUS_PLACEBO variable corresponds to the "Risk Difference" variable in Display 2-3. Note that while you could also have used the BARCHART statement for the Risk Difference and get the same output, typically, BARCHARTPARM is used when the data is pre-summarized and BARCHART is used when the summarization is to take place within the GTL statement.

Output 2-9 shows a bar chart of the risk difference between the xanomeline low dose and placebo sorted in descending risk difference order.

Output 2-9: Bar Chart of Risk Differences

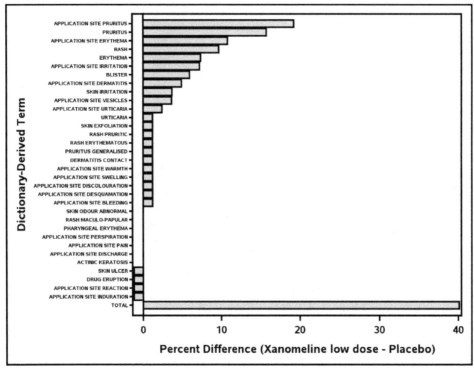

2.8 Adverse Event Profile Plot

AE profile plots are useful to show the adverse events that each patient had, and the duration of each adverse event. Display 2-4 is a data snippet of the profile data for one patient. The data comes from the ADAE analysis data set, and the data snippet contains the unique patient identifier, preferred term, sequence number, relative start day of the AE, and the relative end day.

Display 2-4: Data Snippet of Profile Plot

USUBJID	AEDECOD	AESEQ	ASTDY	AENDY
01-701-1192	APPLICATION SITE ERYTHEMA	13	137	
01-701-1192	APPLICATION SITE IRRITATION	12	137	148
01-701-1192	APPLICATION SITE IRRITATION	15	137	148
01-701-1192	APPLICATION SITE PRURITUS	14	137	
01-701-1192	ERYTHEMA	3	17	55
01-701-1192	ERYTHEMA	6	17	55
01-701-1192	MUSCULAR WEAKNESS	8	48	71
01-701-1192	MUSCULAR WEAKNESS	11	48	71
01-701-1192	NASAL MUCOSA BIOPSY	1	13	13
01-701-1192	PNEUMONIA	7	48	77
01-701-1192	PNEUMONIA	10	48	77
01-701-1192	SECRETION DISCHARGE	2	17	55
01-701-1192	SECRETION DISCHARGE	5	17	55

Program 2-10 shows the code used to produce the AE profile plot. There is a new plot statement in this program called HIGHLOWPLOT. The high-low plot is what is also used to indicate the length of time that the patient had an adverse event.

Program 2-10: Code for AE Profile Plot

```
data adae_trtemfl;
   set adam.adae;
   where TRTEMFL = "Y";
run;

proc sort data = adae_trtemfl;
   by usubjid;
run;

* Template for AE;
proc template;
   define statgraph ae_timeline;
   dynamic _byval_;  ❶
      begingraph;
         discreteattrmap name="Severity";  ❷
            value "MILD" / fillattrs = (color = GraphData3:color)
               lineattrs = (color = GraphData3:contrastcolor pattern = 1)
                        markerattrs = (color = GraphData3:color);
            value "MODERATE" / fillattrs = (color = GraphData1:color)
               lineattrs = (color = GraphData1:contrastcolor pattern = 2)
               markerattrs=(color=GraphData1:color);
            value "SEVERE" / fillattrs = (color = GraphData2:color)
               lineattrs = (color = GraphData2:contrastcolor pattern = 3)
               markerattrs = (color = GraphData2:color);
         enddiscreteattrmap;
         discreteattrvar attrvar = id_severity var = aesev attrmap = "Severity";

         entrytitle halign = center "Unique Patient ID = " _byval_;  ❶
         layout overlay / xaxisopts=(label = "Relative Start Day")
                          griddisplay = on  ❸
                          gridattrs = (thickness = 1px color = lightgray)  ❸
                          linearopts = (minorgrid = true));  ❸
            highlowplot y = aedecod low = astdy high = aendy / type = line
                        highcap = none lowcap = none group = id_severity
                        lineattrs = (thickness = 5px);  ❹
            scatterplot y = aedecod x = astdy /
                        markerattrs = (symbol = trianglerightfilled size = 15)
                        group = id_severity;  ❺
            scatterplot y = aedecod x = aendy /
                        markerattrs = (symbol = triangleleftfilled size = 15)
                        group=id_severity;  ❺
            discretelegend "Severity" / type = line location = outside
                                        valign = bottom across = 3 down = 1
                                        valueattrs = (size = 7pt);     ❻
         endlayout;
      endgraph;
   end;
run;


proc sgrender data = adae_trtemfl template = ae_timeline;
   where usubjid = "01-701-1192";  ❶
   by usubjid;  ❶
run;
```

❶ The DYNAMIC statement and _BYVAL_ is used to indicate that the values that are within the first variable specified in the by statement in the SGRENDER procedure are treated as dynamic values. Therefore, using ENTRYTITLE _BYVAL_, displays the patient's identifier in the plot title. In this instance, the patient's identifier is 01-701-1192.

❷ The DISCRETEATTRMAP defines the colors and line patterns used to distinguish between "MILD", "MODERATE", and "SEVERE" AEs. The colors for "MILD", "MODERATE", and "SEVERE" AEs are green, blue, and red respectively. Those are the colors associated with GraphData3:color, GraphData1:color, and GraphData2:color when using the SAS supplied style CustomSapphire.

❸ The GRIDDISPLAY = ON option turns on the axis grid lines. Turning on the grid lines on the X axis allows for vertical lines to be drawn that correspond to the major tick marks along the X axis. You can further control the appearance of these grid lines, by using the GRIDATTRS option. With the GRIDATTRS option you can make the grid lines a little thinner than the style default and specify that they are a 1px thick, gray line. If you want to see grid lines that occur between the major tick marks, then you can turn on the minor grid lines by setting MINORGRID = TRUE. This option specifies that there should be minor grid lines between the major tick values. You can also turn on grid lines on the Y axis if you use the options in the YAXISOPTS statement.

❹ The HIGHLOWPLOT statement draws a line from the start of the relative AE day to the end of the relative AE day for each AE.

❺ The SCATTERPLOT statements produce triangles, which indicates when the AE starts and ends.

❻ VALUEATTRS option enables you to control the different attributes for the values in the legend. You can specify things such as color and font size.

Output 2-10 shows the AE Profile plot for the patient 01-701-1192.

Output 2-10: AE Profile Plot

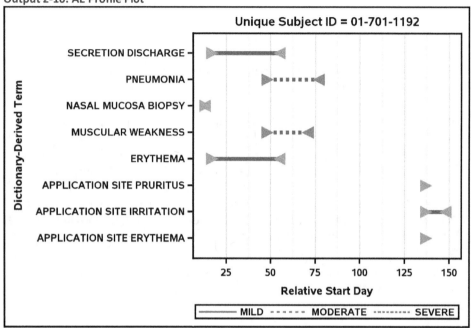

2.9 Chapter Summary

Many pharmaceutical companies adopt the ICH guideline when producing a CSR, and therefore, understanding this guideline can help to identify which plots are needed to support the CSR. Visualization of AEs can help to assess the safety of the investigational product. Bar charts of the number of patients with AEs for each SOC and bar charts of AEs grouped by maximum grade are useful for assessing the patient safety.

Using Discrete Attribute Maps for the Maximum CTCAE Grade is very useful, because this means that there is a one-to-one relationship between the Maximum CTCAE grade and the color it is assigned. So, a Grade 5 CTCAE produces a red bar chart.

Kaplan Meier plots can be useful for assessing adverse events, and these are often requested by regulatory agencies in other countries. Risk Difference bar charts help to easily see the differences in the number of AEs between the treatments. Profile plots are beneficial for an in-depth view of the AEs that each patient had and the AE duration.

References

ICH. (2020, 01 02). (1996) *Structure and Content of Clinical Study Reports.* Retrieved from European Medicines Agency: https://www.ema.europa.eu/en/documents/scientific-guideline/ich-e-3-structure-content-clinical-study-reports-step-5_en.pdf

Kuhfeld, W. F., and So, Y. (2013). "Creating and Customizing the Kaplan-Meier Survival Plot in PROC LIFETEST in the SAS/STAT 13.1 Release." Proceedings of the Annual Pharmaceutical SAS Users Group Conference. Cary, NC: SAS Institute Inc. *PharmaSUG.* Chicago: PharmaSUG.

Kuhfeld, W. F., and So, Y.. (2014). "Creating and Customizing the Kaplan-Meier Survival Plot in PROC LIFETEST in the SAS/STAT® 13.1 Release." Proceedings of the Annual Pharmaceutical SAS Users Group Conference. Cary, NC: SAS Institute Inc. *PharmaSUG.* San Diego: PharmaSUG.

Matange, S. (2013). *Getting Started with the Graph Template Language in SAS®: Examples, Tips, and Techniques for Creating Custom Graphs.* Cary, NC: SAS Institute Inc.

SAS Institute Inc. (2013). *PROC LIFETEST Template (Revised for SAS Global Forum 2013).* Retrieved from SAS Support: http://support.sas.com/documentation/onlinedoc/stat/ex_code/121/templft2.html

SAS Institute Inc. (2019, 12 16). (2013) *PROC LIFETEST Template.* Retrieved from SAS Support: http://support.sas.com/documentation/onlinedoc/stat/ex_code/151/templft.html

Watson, R. (2019). "Great Time to Learn GTL: A Step-by-Step Approach to Creating the Impossible." Proceedings of the *SAS Global Forum 2019 Conference* (p. 1). Cary, NC: SAS Institute Inc.Dallas: SAS Institute.

Watson, R. (2020). "What's Your Favorite Color? Controlling the Appearance of a Graph." Proceedings of the *SAS Global Forum 2020 Conference.* Cary, NC: SAS Institute Inc.

Chapter 3: Using SAS to Generate Graphs for Data Collected Over Time

3.1 Introduction on Findings Data

A big portion of the data collected in a clinical trial is data that is collected over time, such as laboratory (lab) data, vital signs (vitals) data, and quantitative electrocardiogram (ECG) data. This data is used for a variety of things. It can be used to determine whether a patient is eligible for inclusion into a study. In addition, it can help with the diagnosis of a medical condition as well as manage or monitor the patient's condition. These data can also be an indication of an adverse reaction to a treatment. The majority of the time, the data is presented in a tabular format. Due to the copious amounts of parameters that can be collected especially for lab data, viewing the results in a tabular format might not be easy. Because of this, a visual representation of the data can also be provided, and the tabular format can be noted as a reference in case there is a need for further investigation of a particular parameter at a particular time point.

This chapter demonstrates how you can use SAS to produce visualizations on data collected over time in order to help evaluate the safety and efficacy of a study. Although the examples in this chapter are specific to lab data, the types of graphs illustrated can be used for other types of quantitative data that are collected over time.

The ADaM data set from the CDISC SDTM/ADaM Pilot Project used in the examples is the *Analysis Dataset for Laboratory Data* (ADLB).

3.2 Laboratory Values Over Time for Each Laboratory Parameter

During the course of a clinical trial, lab data is collected at various time points to help with the monitoring of the patient's safety as well as to assess the efficacy of medication. Since the data is collected at various time points, one of the key outputs produced is summary statistics for each lab parameter over time. The summary statistics that are produced for each lab parameter can be either for the value at the current time point or for the change from baseline.

3.2.1 Illustrating a Snippet of Laboratory Data

Display 3-1 is a snippet of the data that is used in Program 3-1. ADLB contains a record for each subject (USUBJID), laboratory test (PARAM), and visit (AVISIT). In addition, ADLB contains information such as the corresponding baseline value for each parameter (BASE) so that the change from baseline (CHG) for each post-baseline assessment can be determined. The use of the low and high reference range values (A1LO and A1HI, respectively) helps to determine whether the baseline and analysis value are within the normal reference range. Analysis Reference Range Indicator (ANRIND) is used to determine whether the analysis value (AVAL) is below normal range (L), within normal range (N), or above normal range (H). Baseline Reference Range Indicator (BNRIND) is used to determine whether the corresponding BASE is within the normal reference range.

Display 3-1: Data Snippet of Analysis Data Set for Chemistry Laboratory Data (ADLBCHEM)

Obs	USUBJID	TRTA	AVISIT	ADY	ADT	PARAM	AVAL	BASE	CHG
1	01-701-1015	Placebo	Baseline	-7	26DEC2013	Albumin (g/L)	38	38	.
2	01-701-1015	Placebo	Week 2	15	16JAN2014	Albumin (g/L)	39	38	1
3	01-701-1015	Placebo	Week 4	29	30JAN2014	Albumin (g/L)	38	38	0
4	01-701-1015	Placebo	Week 6	42	12FEB2014	Albumin (g/L)	37	38	-1
5	01-701-1015	Placebo	Week 8	63	05MAR2014	Albumin (g/L)	38	38	0
6	01-701-1015	Placebo	Week 12	84	26MAR2014	Albumin (g/L)	38	38	0
7	01-701-1015	Placebo	Week 16	126	07MAY2014	Albumin (g/L)	37	38	-1
8	01-701-1015	Placebo	Week 20	140	21MAY2014	Albumin (g/L)	37	38	-1
9	01-701-1015	Placebo	Week 24	168	18JUN2014	Albumin (g/L)	38	38	0
10	01-701-1015	Placebo	Week 26	182	02JUL2014	Albumin (g/L)	38	38	0

Obs	A1LO	A1HI	ANRIND	BNRIND
1	33	49	N	N
2	33	49	N	N
3	33	49	N	N
4	33	49	N	N

Obs	A1LO	A1HI	ANRIND	BNRIND
5	33	49	N	N
6	33	49	N	N
7	33	49	N	N
8	33	49	N	N
9	33	49	N	N
10	33	49	N	N

3.2.2 Illustrating Summary Statistics Over Time Represented in Tabular Form

Typically, summary statistics for laboratory data are displayed for each lab parameter, visit, and treatment group. The analysis value and change from baseline can be displayed side by side in the same table as illustrated in Display 3-2. Or some might choose to display a separate table for analysis value and a separate table for the change from baseline. Or some might choose to display the summary statistics as columns and the treatments as rows. Regardless of how the summary statistics are displayed the purpose of the output is similar. You are trying to showcase the summary statistics for each parameter, time point, and treatment in such a way that the results can be compared to assess whether or not a particular treatment is likely to cause an adverse event or if a particular treatment group is efficacious.

Display 3-2: Summary of Change from Baseline Over Time

Laboratory Test	Analysis Visit		Placebo (N=86)		Xanomeline Low Dose (N=84)		Xanomeline High Dose (N=84)	
			Analysis Value	Change from Baseline	Analysis Value	Change from Baseline	Analysis Value	Change from Baseline
Alanine Aminotransferase (U/L)	Baseline	N	86		82		84	
		Mean (SD)	17.6 (9.22)		18.0 (8.72)		19.2 (10.05)	
		Min, Max	7, 69		5, 70		6, 64	
	Week 2	N	83	83	80	78	78	78
		Mean (SD)	18.0 (12.53)	0.2 (7.90)	20.9 (10.55)	2.8 (8.18)	21.0 (8.87)	1.6 (6.83)
		Min, Max	6, 104	-14, 54	5, 88	-43, 22	8, 49	-31, 17
	Week 4	N	79	79	72	70	72	72
		Mean (SD)	18.7 (12.91)	0.7 (8.66)	17.5 (7.66)	-0.7 (5.08)	21.3 (9.51)	2.1 (7.08)
		Min, Max	6, 107	-24, 57	5, 62	-31, 9	8, 61	-28, 21

3.2.3 Illustrating Summary Statistics Over Time Represented in a Graph

Although a table of the summary statistics provides a lot of information, it might not be easy to spot anomalies in the data. However, with a visual representation of the same information, things such as outliers can be easily spotted. Program 3-1 uses ADLBCHEM to produce side-by-side box plots of the change from baseline with a scatter plot overlaid with the individual values grouped by treatment for each parameter and visit. In addition, a table of the summary statistics is displayed below the graph as shown in Output 3-1.

Program 3-1: Box Plots with Scatter Plot Overlay and Summary Statistics Table

```
options nobyline nodate nonumber;
ods graphics on / imagefmt = png width = 5in noborder;   ❶
ods rtf file = 'file location/file name.rtf'
         image_dpi = 300
         style = customSapphire;

title #byval2 at #byval4;   ❷
proc sgplot data = adam.adlbchem noautolegend;   ❸
   where paramcd in ('ALB') and avisitn <= 4;
   by paramcd param avisitn avisit;
   format trtan trt.;

   vbox chg / category = trtan   ❹
              extreme   ❺
              displaystats = (max min std mean n);   ❻

   scatter x = trtan y = chg / jitter;   ❼

   xaxis type = discrete label = "Treatment";   ❽
   yaxis type = linear label = "Change from Baseline";   ❽
run;

ods rtf close;
ods graphics off;
```

❶   By default SAS draws a border around each graph. To remove the border, you can specify NOBORDER or BORDER = OFF or BORDER = NO. All three options will remove the border around the graph.

❷   When a BY statement is used in a procedure, you can insert the BY variable values into the title by using #BYVALn, where n represents the nth variable specified in the BY statement. For example, #BYVAL2 refers to the second by variable, which is PARAM. Therefore, the value of PARAM is inserted into the title. The text "at" is displayed and the value of the 4th by variable, AVISIT, is inserted.

❸   Depending on the type of content, legends are created automatically. However, to turn off this default behavior, you need to indicate NOAUTOLEGEND option in the SGPLOT procedure statement.

❹   To build the box plot, either the VBOX or HBOX statement and the appropriate options are used. The VBOX, as illustrated in Program 3-1, creates a vertical box plot, and HBOX creates a horizontal box plot. The one required argument in the VBOX statement is the numeric analysis variable. If the numeric analysis variable is categorized, then you can specify the variable that is used to categorize the data with the CATEGORY option.

❺   In the box plot statement, there is the concept of "fences." The fences are the location above and below the box as shown in Output 3-1. The upper and lower fences are 1.5 times the distance of the interquartile range (IQR = Q3 − Q1) and the upper far and lower far fences are 3 times the distance of the IQR. Typically, the whiskers on the box plot are drawn from the edge of the IQR to the largest value within the fences. But if you want the whiskers to display to the extreme values even if they are outside the fences, then option EXTREME needs to be specified.

❻ Summary statistics that are associated with the box plots can be displayed below each box using the DISPLAYSTATS option. The default is NONE. However, you can choose to display the STANDARD statistics (that is, N, MEAN, STD), or you can choose ALL available statistics, or you can specify the ones which you want to display. Note that the order in which the statistics are listed in the option statement determines the order in which they are listed in the output. Thus, in order to have "Nobs" print at the top of the list and "Max" to print at the bottom, they need to be listed in reverse order.

❼ The SCATTER statement creates a plot of the individual values in the input data. Within the scatter plot, you can choose whether you want the markers placed on top of each other or if they should be slightly shifted so that it is evident that one marker does not represent multiple values. In order to achieve this effect, the JITTER option should be used. This indicates whether the data markers should be jittered (that is, offset the data markers so that they are not overlaid). JITTER is used when the data has multiple observations that have the same values, typically seen with categorical data. Thus, for this example, the jittering is only done on the X axis because it has a categorical response.

❽ The axes options can be specified using XAXIS and YAXIS if you are using only the bottom and left axes. With the XAXIS and YAXIS statements, you can specify the text that you want displayed and how you want it displayed. In addition, you can indicate what type of axis is used and how the tick marks are displayed.

Output 3-1: Box Plots with Scatter Plot Overlay and Summary Statistics Table

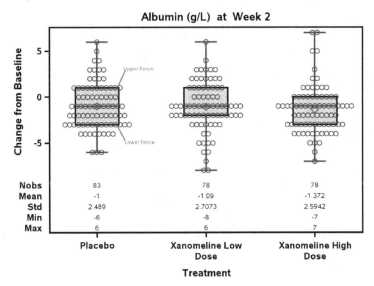

An alternate approach to producing the summary statistics table is to pre-calculate the summary statistics that are needed and merge them onto each corresponding record of the data set that will be used when rendering the template. In the new data set, you can format the statistics so that they are displayed per the desired needs. Display 3-3 shows the augmented ADLBCSTT data set so that it contains the pre-calculated summary statistics in the format that is needed for the display.

- N_C represents the number of patients per treatment stored as a character.

- MEAN_SD is the mean and standard deviation combined into one variable rather than having them on separate rows in the output. By creating MEAN_SD, this allowed for a consistent formatting of the mean and of the standard deviation. For example, all the means are reported to one decimal and the standard deviations are reported to two decimals. Notice in Output 3-1, the format is not consistent.

- MIN_MAX combines the minimum and maximum values into one field so that they can be displayed on one row. The minimum and maximum values are separated by a comma.

With the additional variables that are added into ADLBCSTT, a table can be added to the bottom of the graph. Program 3-2 shows how the XAXISTABLE statement can be used to build the table at the bottom of the graph shown in Output 3-2. In Program 3-2, notice that the DISPLAYSTATS option for BOXPLOT statement is no longer used, instead the XAXISTABLE is used to build the table of summary statistics. Additional features are added to the box plot to help identify the mean, median, and outliers. Furthermore, in Program 3-1, the EXTREME option is used to display the whiskers to the minimum and maximum values; however, in Program 3-2, the EXTREME option is removed so that the whiskers do not extend beyond the IQR. This is to help identify which values are the outliers.

Display 3-3: Data Snippet of Analysis Data Set for Laboratory Data with Summary Statistics Added (ADLBCSTT)

Obs	USUBJID	TRTA	AVISIT	PARAM	CHG	N_C	MEAN_SD	MIN_MAX
1	01-701-1015	Placebo	Week 2	Albumin (g/L)	1	83	-1.0 (2.49)	-6, 6
2	01-701-1023	Placebo	Week 2	Albumin (g/L)	-4	83	-1.0 (2.49)	-6, 6
3	01-701-1047	Placebo	Week 2	Albumin (g/L)	-4	83	-1.0 (2.49)	-6, 6
4	01-701-1118	Placebo	Week 2	Albumin (g/L)	3	83	-1.0 (2.49)	-6, 6
5	01-701-1130	Placebo	Week 2	Albumin (g/L)	1	83	-1.0 (2.49)	-6, 6
6	01-701-1153	Placebo	Week 2	Albumin (g/L)	3	83	-1.0 (2.49)	-6, 6
7	01-701-1203	Placebo	Week 2	Albumin (g/L)	-1	83	-1.0 (2.49)	-6, 6
8	01-701-1234	Placebo	Week 2	Albumin (g/L)	2	83	-1.0 (2.49)	-6, 6
9	01-701-1345	Placebo	Week 2	Albumin (g/L)	-4	83	-1.0 (2.49)	-6, 6
10	01-701-1363	Placebo	Week 2	Albumin (g/L)	-2	83	-1.0 (2.49)	-6, 6

Program 3-2: Box Plots with Scatter Plot Overlay and Summary Statistics Table an Alternate Approach

```
title #byval2 at #byval4;
proc sgplot data = adam.adlbcstt noautolegend;
   where paramcd in ('ALB') and avisitn <= 4;
   by paramcd param avisitn avisit;
   format trtan trt.;

   vbox chg / category = trtan
              nofill  ❶
              meanattrs = (symbol = diamondfilled color = red size = 10) ❷
              medianattrs - (color = green thickness = 4) ❷
              whiskerattrs = (pattern = 2 color = red thickness = 4) ❷
              outlierattrs = (symbol = starfilled color = green size = 12);  ❷

   scatter x = trtan y = chg / jitter;

   xaxistable n_c ❸ / x = trtan  ❹
                   location = inside  ❺
                   separator  ❻
                   valueattrs = (size = 10 ❼
                                 color = cadetblue
                                 weight = bold
                                 style = italic);

   xaxistable ❸ mean_sd min_max / x = trtan  ❹
                                  location = inside;  ❺

   xaxis type = discrete label = "Treatment";
   yaxis type = linear label = "Change from Baseline";
run;
```

❶   If you do not want the boxes to be filled with a color, then you can use the NOFILL option to hide the fill area color.

❷   Within the VBOX statement, you can control the attributes of the different portions of the box plot. For example, you would use the following options to specify the appearance of the specific component of the box plot.

  ○   MEANATTRS enables you to indicate that the symbol is to be a red filled diamond that has a size of 10 pixels.

  ○   MEDIANATTRS enables you to indicate that the median line is a green line that has a thickness of 4 pixels.

  ○   OUTLIERATTRS enables you to identify the outliers as a green filled start that has a size of 12 pixels.

  ○   WHISKERATTRS enables you to specify that the whiskers and cap lines is a red short-dashed line with a thickness of 4 pixels.

  In addition to the ones mentioned, there are other attributes of the box plot that you can govern, such as box outline, the fill area, the fill pattern, and connector lines.

❸   XAXISTABLE enables you to display a table along the horizontal axis. XAXISTABLE enables you to specify what type of information should be displayed, such as summary statistics, and with the variety of options, you have control of how it is displayed. Multiple XAXISTABLE statements can be used within the same PROC SGPLOT statement as illustrated in Program 3-2. The first XAXISTABLE is used to display the number of subjects. The second XAXISTABLE is used to display the mean and standard deviation (StD) as well as the minimum and maximum values. With the use of the XAXISTABLE, you have more control over how the statistics are displayed and what labels are used.

❹   The X option enables you to align the axis table values to the X axis. The value of X must align with the category variable when a categorical plot is used. If the X option is not specified, then the location of the table values is determined based on the X variable for the primary plot.

❺   The axis table can be on the inside of the axis area or the outside. By default, the axis table is located on the outside. To have the table on the inside, you need to use the LOCATION option.

❻   SEPARATOR allows a line to be drawn between the axis table and the preceding plot statement. Notice that in Program 3-2 the SEPARATOR option is only used on the first axis table. This allowed a line to be drawn to separate the plots from the axis table that contained the number of subjects (N_C). If the SEPARATOR option is used on the second XAXISTABLE statement, then a line would have been drawn to separate the number of subjects axis table from the second axis table. Thus, a line would have been drawn above and below the axis table containing the number of subjects. Note that the SEPARATOR only applies when the LOCATION = INSIDE.

❼   In addition, we can control how the value is displayed by using the value attributes (VALUEATTRS). With VALUATTRS, you can specify things such as the size, color, font, weight, and style.

Output 3-2: Box Plots with Scatter Plot Overlay and Summary Statistics Table, an Alternate Approach

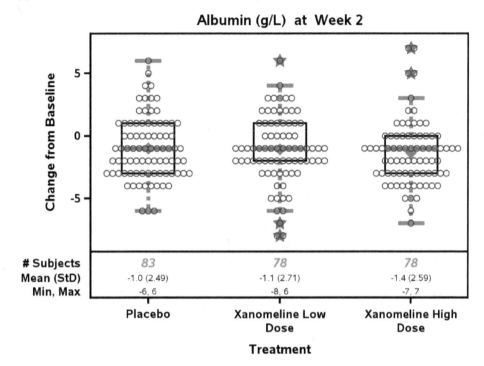

3.3 Individual Patient Changes

Although the change from baseline by lab test and treatment provides a tremendous amount of information about the safety or efficacy of the treatment, it is sometimes necessary to view the data at an individual patient level. Although this information can be viewed through a listing of the patient's values, sometimes a visual representation can be more informative. With a visual representation, you can easily see things such as a spike or a dip in the data.

Program 3-3 is used to produce a series plot of a patient's change from baseline over time for each lab test.

Program 3-3: Series Plot for Change from Baseline Over Time for Each Patient and Laboratory Test

```
title  j = l 'Patient:   ' color = CX3D5AAE height = 11pt '#byval1' ❶
        j = r color = black height = 11pt 'Treatment:  '
            color = CX3D5AAE height = 13pt '#byval2';  ❶
title3 j = l 'Laboratory Test:  ' color = CX3D5AAE height = 11pt '#byval5'; ❶

proc sgplot data = adam.adlbalbp noautolegend;
    by usubjid trtan paramcd param;
    format trtan trt. avisitn vis.;
```

```
    series x = avisitn y = chg / lineattrs = (color = CX90B328 pattern = 4)    ❷
                                  markers    ❸
                                  markerattrs = (symbol = circlefilled    ❸
                                                 color = CX90B328);

    xaxis type = linear label = "Analysis Visit"
          min = 2 max = 26    ❹
          values = (&vlist)    ❺
          fitpolicy =   rotate    ❻
          valuesrotate = diagonal2;    ❼
    yaxis type = linear label = "Change from Baseline" ;
run;
```

❶ Recall from Program 3-1 that the use of #BYVAL*n* on the title statement indicates that the nth BY variable value should be inserted into the title. In addition to specifying the text, you can control the justification and the appearance of the text. With the use of the JUSTIFY (or J) option, you can indicate whether the text should appear left (L), right (R), or center (C) justified. The text that follows the JUSTIFY option is what is justified. Therefore, you can left-justify part of the title and right-justify the other portion. You can use options such as COLOR and HEIGHT to color the appearance of the actual text. Similar to the JUSTIFY option, any text following the option is what is modified. In the example in Program 3-3, in the first title, the first part of the text is left-justified, but only the text associated with #BVYAL1 will have a size of 11 pt and a color associated with the HEX code CX3D5AAE. The portion of the first title that occurs after #BYVAL1 is right-justified, and in order for the indicated text to match the font and size of the text provided in the first portion, you need to change the color to black and the height to 11 pt. The portion associated with #BYVAL2 can then have the color changed to HEX code CX3D5AAE. As demonstrated, you can switch back and forth between colors and height on the same title statement. The color can be specified using one of the seven valid color-naming schemes (Watson, 2020):

   ○   RGB (red green blue), displayed in HEX code

   ○   CMYK (cyan magenta yellow black)

   ○   HLS (hue lightness saturation)

   ○   HSV (hue saturation brightness), also called HSB

   ○   Gray scale

   ○   SAS color names (from the SAS Registry)

   ○   SAS Color Naming System (CNS)

❷ You can create a series plot using the SERIES statement. There are several options that can be used when generating the series plot. One such option is the LINEATTRS. LINEATTRS enables you to specify different attributes like the line pattern, color, and size.

❸ By default, the value markers are not displayed on the series plot. In order to display the markers, the MARKERS option must be specified. You can then further control the appearance of the markers with the MARKERATTRS option.

❹ Within each type of layout there are various options associated with the axes. For Program 3-3, the X axis is linear, so you are able to indicate what the minimum and maximum data value to display on the axis. The MIN and MAX options are used to indicate that only data within that range should be displayed. Any value outside the range is not displayed. Note that this does not indicate the minimum or maximum tick value to display on the axis.

❺ The VALUES indicates the specific values that should be displayed in the order listed. For this output, the macro variable *VLIST* resolves to a space delimited list of visit numbers (that is 2 4 6 8 12 16 20 24 26). Within the SGPLOT procedure a format is applied that allows the visit numbers to be displayed as "Week #" as shown in Output 3-3.

❻ If the text values associated with the visit numbers are used for display in the output, the tick values would have a "collision" if they are displayed using the default options. To avoid this collision, FITPOLICY is used to implement a collision policy. There are a number of possible values for this. However, for this particular output, ROTATE is used to force the tick values to be rotated 45-degree. ROTATE can be used along with VALUESROTATE to indicate how the values are rotated on the axis.

❼ The VALUESROTATE governs the direction of the rotation. If VALUESROTATE is not specified, then ROTATE defaults to the 45-degree diagonal (DIAGONAL). A value of DIAGONAL2 indicates a -45-degree diagonal and a value of VERTICAL is the 90-degree vertical rotation.

Output 3-3: Series Plot for Change from Baseline Over Time for Each Patient and Laboratory Test

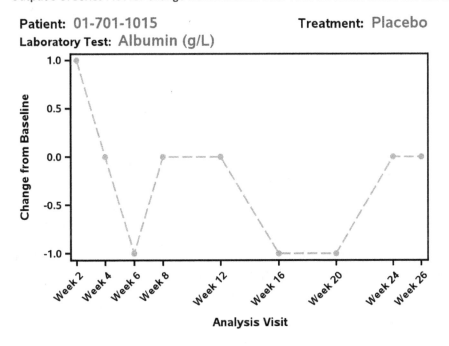

With Program 3-3 it is possible to produce a separate graph for each patient, treatment and lab test combination. Each combination would be its own graph. However, sometimes it is beneficial to see all lab tests that might pertain to certain indications on one page. For example, you might want to monitor the liver function of the patient while they are on medication. Rather than having each test associated with liver function as a separate graph, you can create a graph that will have most if not all associated tests in one output as demonstrated with Program 3-4 which yields Output 3-4.

The major difference is that instead of using the SGPLOT procedure, the SGPANEL procedure is used and a panel variable needs to be specified. The SERIES statement that is in Program 3-3 is exactly the same statement that is used in Program 3-4. However, the axes statements differ slightly to account for the fact that there is more than one graph in the output. For example, PROC SGPLOT uses XAXIS and YAXIS statements, whereas PROC SGPANEL uses COLAXIS and ROWAXIS statements.

Program 3-4: Data Panel of Series Plot for Change from Baseline Over Time for Each Patient

```
title  j = 1 'Patient:  ' color = CX3D5AAE height = 11pt '#byval1';
title2 j = 1 'Treatment:  ' color = CX3D5AAE height = 11pt '#byval2';

proc sgpanel data = adam.adlb3sub noautolegend;   ❶
   by usubjid trtan;
   format trtan trt.;
   panelby param / columns = 4 rows = 2   ❷
                   spacing = 4   ❸
                   novarname   ❹
                   headerattrs = graphdatatext   ❹
                   headerbackcolor = lightgray   ❹
                   skipemptycells;   ❺

   series x = avisitn y = chg / lineattrs = (color = CX90B328 pattern = 4)
                               markers
                               markerattrs = (symbol = circlefilled
                                              color = CX90B328);

   colaxis type = linear label = "Analysis Visit (Week)" ❻
         min = 2 max = 26
         values = (&vlist);

   rowaxis type = linear label = "Change from Baseline" ;  ❻
run;
```

❶ SGPANEL procedure displays a series of graphs that have a similar structure based on the data. The number of cells produced by the PROC SGPANEL is dependent on the number of crossings from the classification variables.

❷ The classification variables are specified using the PANELBY statement. Since there is one classification variable, which is PARAM with seven levels, then only seven cells are created. The number of cells displayed is determined by the number of COLUMNS and ROWS specified.

❸ SPACING option indicates the amount of space that should be between each panel.

❹ The header for each panel generated can be controlled by a variety of options. By default, the variable label is also displayed along with the variable value. For example, the header in the first cell in Output 3-4 would be "Parameter =Albumin (g/L)". However, you can remove the "Parameter = " portion by using the NOVARNAME option. In addition, you can control the background color as well as the font and font size using HEADERBACKCOLOR and HEADERATTRS options. Within the HEADERATTRS options, you can control the various attributes associated with the header such as color and size. Using GRAPHDATATEXT indicates that the smallest font associated with the style should be used. If the header should have a background color so that it stands out, then HEADERBACK can be used to indicate the desired color.

❺ SKIPEMPTYCELLS indicates whether empty cells in a partially filled grid are to be skipped. By default, if the number of crossings from the classification variables does not match the number of columns and rows, then in order to complete the grid the remaining cells that are not filled with data are padded with empty cells. In Output 3-5, there are seven levels for the classification variable, but there are 8 cells (COLUMNS x ROWS = 4 x 2 = 8) that need to be filled. Thus, there is one empty cell. The graph would typically have the X axis for the last column in alignment with the other columns and the 8th cell would have a blank header with a blank cell. However, with the use of SKIPEMPTYCELLS, the empty cells are removed, and the X axis is moved below the data in column 4 row 1.

❻ The axes options specified in PROC SGPLOT in Program 3-3 are now specified in the PROC SGPANEL using the COLAXIS and ROWAXIS statements. Notice that for this particular program the FITPOLICY and VALUESROTATE options are no longer used since the visits are no longer being formatted to "Week #". Only the numeric value of AVISITN is displayed on the X axis (COLAXIS) of the graph.

Output 3-4: Data Panel of Series Plot for Change from Baseline Over Time for Each Patient

Patient: 01-701-1015
Treatment: Placebo

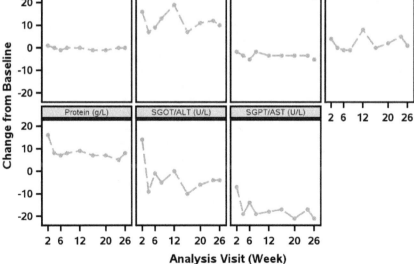

3.4 Mean Change from Baseline Over Time

In Section 3.2 *Illustrating Laboratory Values Over Time for Each Laboratory Parameter*, the use of box plots with scatter plots overlaid are generated for each combination of lab test and visit. Thus, creating separate plots for each combination. However, it is sometimes necessary to see how the values are changing over time and have these values displayed in one graph. So, although this information can also be viewed in table format as shown in Display 3-2, the visual representation helps draw the reviewer to the data that might be cause for concern. Program 3-5 shows how a series plot of the mean change from baseline over time can be produced and have all treatments displayed on the same graph.

In order to produce the mean change from baseline over time with "whiskers" to indicate +/- 1 standard error from the mean, it might be necessary to do some pre-processing to obtain the summary statistics. Display 3-4 shows a snippet of the data after the summary statistics have been generated.

- CHG_N: Number of records with nonmissing change from baseline value within the parameter, treatment, and visit

- CHG_MEAN: Mean of all nonmissing change from baseline values within the parameter, treatment, and visit

- CHG_STDERR: Standard error associated with CHG_MEAN

- LERR: CHG_MEAN – CHG_STDERR

- UERR: CHG_MEAN + CHG_STDERR

This data is used in Program 3-5 in order to produce Output 3-5. Note that it is not required that you use +/- 1 standard error. You could use +/- 2 standard deviations or lower and upper limit of 95% confidence interval. The determination of the lower and upper "whisker" end point is based on client needs; however, the concept to display them is the same regardless of what is used for "whisker" end points.

Display 3-4: Data Snippet of Summary Statistics for Each Lab Test, Treatment, and Visit (ADLBSTAT)

Obs	PARAM	AVISIT	TRTA	CHG_N	CHG_MEAN	CHG_STDERR	LERR	UERR
1	Albumin (g/L)	Week 2	Placebo	83	-1.00000	0.27320	-1.27320	-0.72680
2	Albumin (g/L)	Week 2	Xanomeline Low Dose	78	-1.08974	0.30654	-1.39629	-0.78320
3	Albumin (g/L)	Week 2	Xanomeline High Dose	78	-1.37179	0.29374	-1.66554	-1.07805
4	Albumin (g/L)	Week 4	Placebo	79	-1.01266	0.30227	-1.31492	-0.71039
5	Albumin (g/L)	Week 4	Xanomeline Low Dose	70	-1.18571	0.31740	-1.50311	-0.86831
6	Albumin (g/L)	Week 4	Xanomeline High Dose	72	-1.30556	0.31872	-1.62427	-0.98684
7	Albumin (g/L)	Week 6	Placebo	73	-0.97260	0.26386	-1.23647	-0.70874
8	Albumin (g/L)	Week 6	Xanomeline Low Dose	60	-1.20000	0.32137	-1.52137	-0.87863
9	Albumin (g/L)	Week 6	Xanomeline High Dose	66	-0.96970	0.31951	-1.28920	-0.65019
10	Albumin (g/L)	Week 8	Placebo	72	-0.40278	0.31873	-0.72151	-0.08405

Program 3-5: Mean Change from Baseline Over Time by Treatment

```
proc fontreg mode = all msglevel=verbose; ❶
   fontpath "font location\primetime";
run;

title j = c 'Laboratory Test: 'font = 'primetime' ❶ bold height = 11pt
'#byval2';

proc sgplot data = adam.adlbstat;
   by paramcd param;
   format trtan trt. avisitn vis.;

   series x = avisitn y = chg_mean / markers
                                 group = trta; ❷

   scatter x = avisitn y = chg_mean / group = trtan  ❷
                               yerrorlower = lerr  ❸
                               yerrorupper = uerr; ❸

   xaxis type = linear label = "Analysis Visit"
         min = 2 max = 26
         values = (&vlist)
         fitpolicy =  rotate
         valuesrotate = diagonal2;
   yaxis type = linear label = "Change from Baseline" ;
run;
```

❶ One of the options that can be used to control the appearance of text is the FONT option. The font can be one of the fonts that are part of the SAS registry or it can be a font that you add to the SAS registry so that it can be used in various SAS outputs. For information about how to read in a font, visit the FONTREG Procedure in the Base SAS® 9.4 Procedures Guide, Seventh Edition documentation (SAS Institute Inc., 2020). For Output 3-5, a custom font name "primetime" is used to display the lab test name. Custom fonts can be company-specific fonts or in the case of "primetime," it can be found online at dafont.com (PRIMETIME, n.d.)

❷ For both the SERIES and SCATTER statements, the GROUP option can be used to create separate series plots and scatter plots for each group specified, that is each unique treatment. If no additional options are indicated to specify how the markers and lines should look for each treatment, then by default the GraphDataN pre-defined style associates the markers and line attributes with the style that is used.

❸ With the use of YERRORLOWER and YERRORUPPER, you can draw the "whiskers" to indicate +/- one standard error from the mean. With the use of this, not only do you see how the means are changing over time but you can see the spread of the means for each visit.

Output 3-5: Mean Change from Baseline Over Time by Treatment

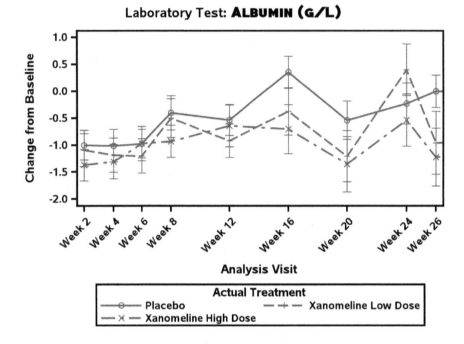

3.5 Laboratory Values Shift from Baseline to Post-Baseline Plot

As previously mentioned, the most common way to display lab data is in a tabular form. Another such output is a shift from baseline to maximum post-baseline. With all lab tests there are normal reference ranges indicating if the lab test value is below the normal range, within the normal range, or above the normal range. The ranges are typically set by the lab that is running the specimen to obtain the lab results.

3.5.1 Illustrating Shift from Baseline to Maximum Post-Baseline Normal Range Indicator in Tabular Form

Display 3-5 illustrates a sample of what a possible shift from baseline to maximum post-baseline based on the normal range indicator might look like. The rows and columns of the table are dependent on the client's needs, so it is not uncommon to have additional information such as clinically low or degree below normal range, clinically high or degree above normal range or missing added to such a table. Regardless of the number of categories that are displayed in the table, the purpose is the same; you are trying to determine how many patients shifted to a different level from baseline to post-baseline and if there is a high number of shifts, that could be an indication of a safety issue or an effective treatment.

In order to obtain a shift from baseline table, the baseline value reference range indicator (BNRIND) and the current analysis value reference range indicator (ANRIND) are used to determine the number of subjects in each crossing of BNRIND and ANRIND. These range indicators are determined based on the reference range that is provided by the lab, A1LO for low normal range and A1HI for the high normal range. Thus, the same data in Display 3-1 is used and processed using the FREQ procedure to obtain the counts for BNRIND crossed with ANRIND.

The concepts for assessing the shift from baseline to maximum post-baseline can also be applied to lab tests that use toxicity grading. Thus, a shift from baseline toxicity to maximum post-baseline toxicity can be displayed in a similar manner as Display 3-5.

Display 3-5: Shift from Maximum Baseline to Maximum Post-Baseline Normal Range Indicator

Laboratory Test	Maximum Baseline Result	Maximum Post-Baseline Result		
		Low	Normal	High
Albumin (g/L)	Low	0 (0.0%)	0 (0.0%)	0 (0.0%)
	Normal	1 (1.2%)	83 (96.5%)	0 (0.0%)
	High	0 (0.0%)	0 (0.0%)	0 (0.0%)
Alk. Phos. (U/L)	Low	0 (0.0%)	0 (0.0%)	0 (0.0%)
	Normal	0 (0.0%)	76 (88.4%)	6 (7%)
	High	0 (0.0%)	0 (0.0%)	2 (2.3%)
Bilirubin (umol/L)	Low	0 (0.0%)	0 (0.0%)	0 (0.0%)
	Normal	0 (0.0%)	78 (90.7%)	6 (7%)
	High	0 (0.0%)	0 (0.0%)	0 (0.0%)

3.5.2 Illustrating Shift from Baseline to Maximum Post-Baseline Result in Tabular Form

Although a shift table is informative, it can be easy to overlook some key data points due to the amount of lab tests contained in the table. A supplemental output to the shift table would be to create a shift plot that is a visual representation of the table. With the shift plot, you can showcase the baseline result as well as the maximum post-baseline result.

Because lab tests can be done at a local lab or at a central lab, the reference ranges might be different from patient to patient. If both local and central labs are used, or if more than one central lab is used, the method of standardizing the data so that it can be summarized together is dependent on client-specific rules. For the purpose of this chapter, we are creating a new low and high normal range to standardize across all subjects for each given parameter, and resetting

BNRIND and ANRIND as shown in Display 3-6. The data in Display 3-6 is used in Program 3-6 in order to produce Output 3-6.

- BASE: Baseline Value
- BNRIND: Baseline Reference Range Indicator
- MAXPOST: Maximum Analysis Value Post-Baseline
- MINA1LO: Minimum Analysis Lower Limit
- MAXA1HI: Maximum Analysis Upper Limit
- ANRIND: Analysis Reference Range Indicator

Display 3-6: Data Snippet of Maximum Post-Baseline Values with Standardized Normal Ranges (ADLBMAX)

Obs	TRTA	PARAM	USUBJID	BASE	BNRIND	MAXPOST	MINA1LO	MAXA1HI	ANRIND
1	Placebo	Albumin (g/L)	01-701-1015	38	N	39	33	49	N
2	Placebo	Albumin (g/L)	01-701-1023	40	N	36	33	49	N
3	Placebo	Albumin (g/L)	01-701-1047	42	N	38	33	49	N
4	Placebo	Albumin (g/L)	01-701-1118	39	N	44	33	49	N
5	Placebo	Albumin (g/L)	01-701-1130	37	N	39	33	49	N
6	Placebo	Albumin (g/L)	01-701-1153	39	N	42	33	49	N
7	Placebo	Albumin (g/L)	01-701-1203	38	N	43	33	49	N
8	Placebo	Albumin (g/L)	01-701-1234	41	N	43	33	49	N
9	Placebo	Albumin (g/L)	01-701-1345	45	N	47	33	49	N
10	Placebo	Albumin (g/L)	01-701-1363	41	N	42	33	49	N

Program 3-6: Shift from Baseline to Maximum Post-Baseline Result in Graph

```
title j = c h = 11pt 'Laboratory Test: ' color = green italic bold '#byval2';

proc sgplot data = adam.adlbmax noautolegend;
   by paramcd param;
   format trtan trt.;

   scatter x = base y = maxpost / group = trta name = 'scat';   ❶

   lineparm x = mina1lo y = mina1lo slope = 1;   ❷

   refline mina1lo / axis = x lineattrs = (pattern = 2 color = red);   ❸
   refline maxa1hi / axis = x lineattrs = (pattern = 3 color = cornflowerblue);   ❸
   refline mina1lo / axis = y lineattrs = (pattern = 2 color = red);   ❸
   refline maxa1hi / axis = y lineattrs = (pattern = 3 color = cornflowerblue);   ❸

   text x = mina1lo y = mina1lo text = minlbl / textattrs = (weight = bold   ❹
                                                 color = red)
                                      position = topleft;
```

```
       text x = maxa1hi y = maxa1hi text = maxlbl / textattrs = (weight = bold ❹
                                                               color = cornflowerblue)
                                                    position = topleft;
   keylegend 'scat' / across = 1 ❺
                      border
                      location = outside
                      position = bottomleft
                      title = 'Actual Treatment';

   legenditem type = line name = 'LLN' / label = 'Lower Range' ❻
                                         lineattrs = (pattern = 2
                                                      color = red);
   legenditem type = line name = 'ULN' / label = 'Upper Range' ❻
                                         lineattrs = (pattern = 3
                                                      color = cornflowerblue);
   keylegend 'LLN' 'ULN' / across = 1 ❺
                           border
                           location = outside
                           position = bottomright
                           title = 'Normal Reference Range';

   xaxis type = linear label = "Maximum Baseline Results";
   yaxis type = linear label = "Maximum Post Baseline Result";
run;
```

❶ Although Chapter 2 used legends, we go a bit into how those legends are created. In order for a legend to be created using information from another plot statement within the procedure, the plot statement must be provided a NAME so that it can be referenced by other statements within the procedure. The NAME can be used in the KEYLEGEND statement to build the legend or in this case legends. Note that the casing between the NAME indicated on the plot statement that is to be referenced must match on the plot statement where it is referenced.

❷ LINEPARM is used to create the diagonal line in Output 3-6. The diagonal line represents no change. In order to draw this line in Output 3-6, you need to specify the X coordinate and the Y coordinate to indicate the starting point as well as the SLOPE. In this illustration, the starting point is at the low normal range (MINA1LO) on the X and Y axes and set the slope to 1. Note that SLOPE can be set to a positive or negative value. The value determines the angle of the line. Positive slopes up to the right and negative slopes down to the right. If SLOPE is set to 0, then a horizontal line parallel to the X axis is drawn, while a SLOPE that is missing draws a vertical line parallel to the Y axis.

❸ If you know you want to create a horizontal or vertical line, rather than using LINEPARM, you can use REFLINE to draw the line. When using REFLINE, you need to indicate the variable or specific value(s) that the reference line is based on. If a variable is specified, then a reference line for each unique value of that variable will be drawn. Either the reference line variable or specific value(s) is a required argument, but you can have additional arguments such as AXIS to specify which axes the reference line should be drawn on. Like most of the other plot statements, there are various options that will control the appearance of the line. For a complete list of line patterns, visit Graph Template Language User's Guide: Available Line Patterns in the SAS® 9.4 and SAS® Viya® 3.2 Programming Documentation (SAS Institute Inc. 2020).

❹ To display text at a specific location within the graph, you can use the TEXT statement. This enables you to put the text associated with the variable indicated by the TEXT argument at the (X, Y) location. The appearance of the text can be controlled by the TEXTATTRS options. In addition, you can specify where the text should appear with respect to the data point that is represented by the (X, Y) coordinate. For example, should the text appear above (TOP), below (BOTTOM), immediately to the left (LEFT), or slightly above and to the left (TOPLEFT)? Additional locations are TOPRIGHT, RIGHT, CENTER, BOTTOMLEFT, BOTTOMRIGHT. The default value is CENTER.

❺ KEYLEGEND is used to add a legend to a plot so that you can better understand the lines and markers that are displayed. Various options are valuable that govern how the legend looks. With the ACROSS option, you can specify how many legend entries should be displayed horizontally before moving to the next row. In addition, you can add a border around the legend by using the BORDER option. Note that the default options specifies BORDER, so it is not necessary to actually specify the option. To place the legend on the inside or the outside of the graph, you need to indicate INSIDE or OUTSIDE on the LOCATION option. You can provide further instructions on where on the output the legend should be. Should it be at the BOTTOM or TOP? The POSITION option enables you to have this control. Being able to specify the LOCATION and POSITION can be beneficial especially if you have multiple KEYLEGEND statements. If multiple KEYLEGEND statements are used, then by default each legend is placed in a different location to avoid a "collision." But since it is possible to overwrite the default behavior and specify the location of each legend, you need to take care to make sure they do not collide. With the use of multiple KEYLEGEND, it might be beneficial to label each legend for clarity. A title can be added to each legend using the TITLE option.

❻ There might be times when an item is needed for a legend that is not readily available from one of the other plot statements. If this situation arises, you can add an item using the LEGENDITEM statement. In Output 3-6, there are normal reference range lines. You want to add this to the legend in case the "LLN" and the "ULN" that are on the diagonal where the corresponding reference lines intersection are not discernible or if they are omitted. You want to ensure that the individual looking at the graph understands what those reference lines represent. In order to add an item to the legend, you need to indicate the TYPE, which represents the item on the graph, such as a line or a marker. Possible values for TYPE are LINE, MARKER, MARKERLINE, TEXT, or FILL. You also need to provide the LEGENDITEM a NAME, so that it can be referenced in another plot statement within the procedure. Additional options can be used to control the appearance and what is displayed as a description for the item (LABEL).

Output 3-6: Shift from Baseline to Maximum Post-Baseline Result in Graph

Laboratory Test: *Albumin (g/L)*

3.6 Chapter Summary

Because laboratory data provides so much insight into a clinical trial, it is important to be able to display the data in such a way that it makes it easy to monitor the safety of patients as well as the efficacy of the treatment. Although the majority of the output used is in a table format, it is beneficial to also have a visual representation of the output. The visual representation makes it easier to spot the data that needs further investigation.

While this chapter focused on outputs that pertain to lab data, these same graphs and concepts can be applied to other types of data that are collected over time, such as vitals and ECG data.

References

Matange, S. (2013). *Getting Started with the Graph Template Language in SAS®: Examples, Tips, and Techniques for Creating Custom Graphs.* Cary, NC: SAS Institute Inc.

PRIMETIME. (n.d.). Retrieved from dafont.com: https://www.dafont.com/primetime.font

SAS Institute Inc. (2020, August 20). Base SAS Procedures Guide: *FONTREG Procedure.* Retrieved from SAS® 9.4 and SAS® Viya® 3.5 Programming Documentation: https://documentation.sas.com/?docsetId=proc&docsetTarget=p199jutvq45hmun1tg4h11umqnsw.htm&docsetVersion=9.4&locale=en

SAS Institute Inc. (2020, July 30). SAS *Graph Template Language: User's Guide: Available Line Patterns.* Retrieved from SAS® 9.4 and SAS® Viya® 3.2 Programming Documentation: https://documentation.sas.com/?cdcId=pgmsascdc&cdcVersion=9.4_3.2&docsetId=grstatug&docsetTarget=n13pm0ndse66l2n1u309543mx2yt.htm&locale=en

Watson, R. (2019). "Great Time to Learn GTL: A Step-by-Step Approach to Creating the Impossible." Proceedings of the *SAS Global Forum 2019 Conference.* Cary, NC: SAS Institute Inc. (p. 1). Dallas: SAS Institute.

Watson, R. (2020). "What's Your Favorite Color? Controlling the Appearance of a Graph." Proceedings of the *SAS Global Forum 2020 Conference.* Cary, NC: SAS Institute Inc.

Chapter 4: Using SAS to Generate Graphs for Exposure and Patient Disposition

4.1 Introduction to the Napoleon Plot/Swimmer Plot

The Napoleon Plot, also known as the swimmer plot, is useful for assessing the safety of a treatment. With a Napoleon Plot, you can convey a variety of information either from one data source or multiple data sources in one graph. As mentioned in Harris (2014), another way to use the Napoleon Plot is to use it to assess the Maximum Tolerated Dose (MTD) exclusively. That is, display the adverse events on the plot that contribute to determining if the MTD has been met. There are several different scenarios in which a Napoleon Plot can be useful. By putting the various pieces of data for each patient on one plot, you can better understand the patient's safety data. This chapter focuses on the exposure of the treatment and the possible reasons for treatment discontinuation.

The plot that conveys a variety of information relating to the safety of the treatment is called a Napoleon Plot because the name is meant to imply that several different types of data are all displayed conveniently on one single plot. The famous plot of Napoleon's march, for example, showed things like troop movements, temperature, etc. In this context, dose, time on study, number of cycles, and reason for discontinuation are conveyed all in one image. Similarly, the reason why the plot is referred to as the swimmer plot is because when you look at all the patients' time on study, it looks like a swimmer lane.

4.2 Creating a Napoleon Plot/Swimmer Plot

There are a few ways to produce the horizontal lines in the Napoleon plot, which generally denote the treatment duration. You can achieve this by using either the BARCHARTPARM, HIGHLOW, or VECTORPLOT statements. This chapter will first focus on using the

BARCHARTPARM statement to indicate the patient duration and second focus on using the HIGHLOW statement to indicate patient duration and dose interruption. For an illustration of the VECTORPLOT statement, refer to Section 6.6 *Swimmer Plot for Overall Response During Course of Study*, where it demonstrates an example using the HIGHLOW statement to show the treatment duration, while using the VECTORPLOT statement to highlight the tumor response duration. Using the HIGHLOW statement makes it easier to show treatment dose interruptions than using the BARCHARTPARM statement. Other relevant statements for the creation of the Napoleon Plot are the SCATTERPLOT statement for displaying symbols, such as the reason for treatment discontinuation and the DISCRETELEGEND statement for displaying the legend. Generally, the GROUP statement is used with Napoleon Plots. Therefore, to ensure that the groups have the desired attributes, even when a group is missing, using attribute maps or an attribute map data set is very useful. You can create and reference attribute maps using the DISCRETEATTRMAP and DISCRETEATTRVAR statements as shown in Section 2.5.5 *Producing Frequency Plot of Adverse Events Grouped by Maximum Grade for Each System Organ Class*.

Due to the complexity of creating a Napoleon Plot, the process is broken down into four steps. Each step builds upon the other in order to produce the final plot. The four steps for creating the Napoleon Plot are:

1. Creating the bar charts and dose legend.
2. Creating reason for treatment discontinuation markers and legend.
3. Adding treatment discontinuation initialisms to the plot.
4. Adding cohort information to the plot.

The data used in the Napoleon Plot can be extracted from the SDTM domains: DS, EX, SV, and DM. Display 4-1 shows a data snippet of the data used in the Napoleon Plot. The variables in the data snippet are:

- SUBJID: the patient identifier.

- ORDER_SUBJECT: a variable that indicates the order the patients should follow on the graph. Patients with the longest treatment exposure are ordered first.

- CYCLE_DAYS: indicates the number of days that there are in a cycle. Generally, there are 28 days in a treatment cycle. Multiple cycle days indicate multiple cycles. For example, patient 1003 had four cycles.

- LENGTH: indicates the total number of days that the patient has currently been on the treatment. For instance, patient 1003 has four cycles each lasting 28 days, so patient 1003 has been on treatment for 4 * 28 = 112 days.

- DOSE: the dose the patient has received at each visit. You can see that patient 1004 received 200 mg of treatment at Visit 1. However, this was reduced to 100 mg at Visit 2, and this was further reduced to 50 mg at Visits 3 and 4.

- VISIT_DOSE: concatenates the VISIT and DOSE variables. Although the VISIT variable is not shown on this data snippet in order to save space. The VISIT_DOSE variable is useful for identifying when a dose has changed, and displaying that change on the graph.

- RESULT: contains the initial(s) of the treatment discontinuation reason. "O" means that patient is still "Ongoing" and has not discontinued treatment yet. "AE" stands for "Adverse Event."

- RESULT_LABEL: shows the initial(s) of the treatment discontinuation for one visit per patient. It is useful for adding the initial onto the graph.

- COHORT_LABEL_X: specifies the X-axis coordinate of the cohort label. COHORT_LABEL_X is equal to the LENGTH variable + 15.

- COHORT_LABEL: shows the patients' cohort. This is useful for displaying the cohort on the graph. Notice that the variable is only populated on the first record for each patient. This is because it will only need to be displayed once as illustrated in Output 4-4.

Display 4-1: Data Snippet Used for Napoleon Plots (NAPOLEON_DATA1)

Obs	SUBJID	ORDER_SUBJECT	CYCLEDAYS	LENGTH	DOSE	VISIT_DOSE	RESULT	RESULT_LABEL	COHORT_LABEL_X	COHORT_LABEL
1	1003	1	28	112	200	1-200	O	O	127	Cohort 1
2	1003	1	28	112	200	2-200	O		127	
3	1003	1	28	112	200	3-200	O		127	
4	1003	1	28	112	200	4-200	O		127	
5	1004	2	28	112	200	1-200	AE	AE	127	Cohort 1
6	1004	2	28	112	100	2-100	AE		127	
7	1004	2	28	112	50	3-50	AE		127	
8	1004	2	28	112	50	4-50	AE		127	

4.2.1 Creating the Bar Charts and Dose Legend

One of the desired outcomes is to be able to determine whether the patient had a dose reduction. In order to clearly identify patients who had their dose reduced and at which particular visit, the bar charts need to be stacked up at each visit, where each visit shows the number of days that the subject is on treatment, as illustrated in Output 4-1. Therefore, the data should be in a structure to allow the bar charts to be stacked up, as shown in Display 4-1.

The structure of the data in Display 4-1 also enables you to produce the graph and display all the key information. As shown in Output 4-1, which is produced by Program 4-1, you can view the number of cycles and the number of days the patient was on treatment for each cycle. Thus, by having the data presented in this fashion, you can easily see whether the patient maintained their dosage or whether they reduced their dosage.

Program 4-1: Code for Creating Stacked Bar Charts and Dose Legend

```
proc template;
   define statgraph napplot1;
      begingraph;

         discreteattrmap name = "Dose_Group"; ❶
            value "50" / fillattrs = (color = white)  ❶
                         lineattrs = (color = black pattern = solid);
            value "100" / fillattrs = (color = yellow transparency = 0.3) ❶
                          lineattrs = (color = black pattern = solid);
            value "200" / fillattrs = (color = darkbrown) ❶
                          lineattrs = (color = black pattern = solid);
         enddiscreteattrmap;
         discreteattrvar attrvar = id_dose_group var = dose attrmap = "Dose_Group"; ❷

         discreteattrmap name = "Visit_Dose_Group"; ❸
            value "1-50" "2-50" "3-50" "4-50" /
                    fillattrs = (color = white)
                    lineattrs = (color = black pattern = solid);
            value "1-100" "2-100" "3-100" "4-100" /
                    fillattrs = (color = yellow transparency = 0.3)
                    lineattrs=(color = black pattern = solid);
            value "1-200" "2-200" "3-200" "4-200" /
                    fillattrs = (color = darkbrown)
                    lineattrs = (color = black pattern = solid);
         enddiscreteattrmap;
         discreteattrvar attrvar = id_visit_dose_group
                         var = visit_dose
                         attrmap = "Visit_Dose_Group";

         entrytitle "Napoleon Plot" / textattrs = (size = 14px);
         layout overlay / xaxisopts = (label = "Days on Treatment")

                          yaxisopts = (label = "Patient"
                                       reverse = true);  ❹

            barchartparm x = order_subject y = cycledays / ❺
               barwidth = 0.8 ❺
               orient = horizontal ❺
               group = id_visit_dose_group ❺

            discretelegend "Dose_Group"/ type = fill ❻
                                         autoalign = (bottomright) ❻
                                         across = 1 ❻
                                         location = inside ❻
                                         title = "Dose"; ❻

         endlayout;
      endgraph;
   end;
run;

proc sgrender data = tfldata.napoleon_data1 template = napplot1;
   format order_subject subjfmt.;  ❼
run;
```

❶ DISCRETEATTRMAP and the VALUE statements are used to define the attributes for each Dose Group. For example, the attributes for the 50 mg dose group are defined as having a white fill color with a solid black outline. A VALUE statement is needed for each unique value so that a specific color scheme can be used for that value. The specified attributes means that for a bar chart, as shown in Output 4-1, for the "50" mg dose group will have a black outline color, and the fill color of the bar will be white.

❷ DISCRETEATTRVAR is used to associate the attributes that are defined for each value in the DISCRETEATTRMAP section with the variable in the data set, which contains those values. In the NAPOLEON_DATA1 data set, the DOSE variable contains the 50, 100, and 200 values. DISCRETEATTRVAR creates an association between the attribute map DOSE_GROUP and the DOSE variable from the data set. This association is called ID_DOSE_GROUP. Therefore, when ID_DOSE_GROUP is referenced in a plot or legend statement, the correct color scheme will be used for the different values of DOSE.

❸ An important aspect of the Napoleon Plot when showing the patient's exposure is highlighting when the patient has reduced their dosage. A patient's dosage might be reduced if it is deemed safer for the patient to be on a lower dose. The attribute map "VISIT_DOSE_GROUP" can help to highlight dose reductions by displaying different attributes for a patient between visits. In essence, each dose is denoted by specific attributes regardless of the visit and hence is the same color that was specified for the overall dose value that is indicated in the first DISCRETEATTRMAP defined for this template. For example, if a patient received 50 mg at visits 1, 2, 3, or 4, then this patient would have a white filled bar chart with a solid black outline for each of the visits that the patient had received 50 mg.

❹ The REVERSE = TRUE option enables the patients on the Y axis to be displayed as expected, with the patients that have been on treatment the longest appearing at the top of the Y axis.

❺ The BARCHARTPARM statement plots the bar chart showing the dose the treatment had at each visit, and for how long the patient was on treatment. Note that while you could also have used the BARCHART statement for the Napoleon Plot and get the same output, typically, BARCHARTPARM is used when the data is pre-summarized and BARCHART is used when the summarization is to take place within the GTL statement.

- The BARWIDTH = 0.8 option makes the width of the bar charts a fraction smaller than their default width of 0.85. You can choose between a minimum width of 0 and a maximum width of 1.

- In the BARCHARTPARM statement, the patient is assigned to the X (or CATEGORY) argument, and the days on treatment for each cycle are assigned to the Y (or RESPONSE) argument. In the BARCHARTPARM statement, X and CATEGORY can be used interchangeably and Y and RESPONSE can be used interchangeably. The ORIENT option specifies the orientation of the bars, and the default orientation option is VERTICAL. Using the default option specifies that the category variable appears on the X axis, and the response variable appears on the Y axis. In Program 4-1, the ORIENT = HORIZONTAL option specifies that the category variable, that is, the patient identifier, appears on the Y axis, and the response variable, that is the days on treatment appears on the X axis.

- The GROUP = ID_VISIT_DOSE_GROUP option assigns the colors of the bar charts based on the dose the patient had at that visit. ID_VISIT_DOSE_GROUP is the name of the association between the VISIT_DOSE_GROUP attribute map and the variable that is defined in the DISCRETEATTRVAR statement.

❻ The DISCRETELEGEND statement plots the dose legend.

- The TYPE = FILLCOLOR adds a square filled in color to represent the dose color. If you did not specify a TYPE, then SAS would use the default ALL option, and a legend will not be produced. The legend will not be created because the legend is based on a bar chart and no markers were used. Similarly, using TYPE = MARKER, TYPE = MARKERCOLOR or TYPE = MARKERSYMBOL will not produce a legend in these circumstances.

- The AUTOALIGN = (BOTTOMRIGHT) option, positions the legend in the bottom right-hand corner.

- The ACROSS = 1 option makes the legend one column wide, so all the three doses are displayed in one column.

- The LOCATION = INSIDE option, positions the legend within the wall area.

- The TITLE = "Dose" option provides a title for the legend and therefore adds the "Dose" title above the three doses.

❼ The FORMAT statement displays the patient number on the Y axis, instead of the values of the variable that represented the ordered patient number.

Output 4-1 shows how long each patient was on treatment for and the dose that the patient received at each cycle.

Output 4-1: Plot of Stacked Bar Charts and Dose Legend

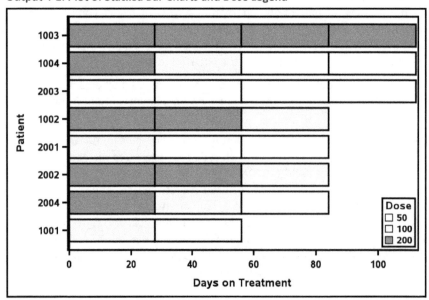

4.2.2 Creating Reason for Treatment Discontinuation Markers and Legend

Program 4-1 showed you how you could create the stacked bar chart and dose legend. Program 4-2 will show you how to add to Program 4-1 and also display the reason for treatment discontinuation markers and reason for treatment discontinuation legend. Program 4-2 will only show the additional code that is needed, and will not show the entire code that is actually used. However, the entire code for all the programs in this chapter is available on GitHub at https://github.com/rwatson724/SAS-Graphs-Clinical-Trials-Example.

Program 4-2: Code for Creating Reason for Treatment Discontinuation Markers and Legend

```
proc template;
   define statgraph napplot2;
      begingraph;

         << SAS CODE for attribute maps >>

         discreteattrmap name = "Response_Markers";  ❶
            value "AE" "Adverse Event" /  ❶
               markerattrs = (color = GraphData4:color symbol = squarefilled);
            value "D" "Death" /  ❶
               markerattrs = (color = GraphData5:color symbol = squarefilled);
            value "O" "Ongoing" /  ❶
               markerattrs = (color = GraphData6:color symbol = squarefilled);
            value "PD" "Progressive Disease" /  ❶
               markerattrs = (color = GraphData7:color symbol = squarefilled);
         enddiscreteattrmap;
         discreteattrvar attrvar = id_response_markers
                     var = result
                     attrmap = "Response_Markers";

         << SAS CODE for layout and bar chart statements >>

            scatterplot x = length y = order_subject /  ❷
               markerattrs = (size = 12pt)
               group = id_response_markers;


         << SAS CODE for "Dose" discrete legend >>

            discretelegend "Response_Markers"/ type = marker  ❸
                                               exclude = ("AE" "D" "O" "PD")  ❸
                                               location = outside  ❸
                                               valign = bottom  ❸
                                               down = 1   ❸
                                               title = "Reason for Treatment
                                                  Discontinuation"  ❸
                                               Titleattrs = (size = 8pt)  ❸
                                               Valueattrs = (size = 7pt);  ❸
      endgraph;
   end;
run;

proc sgrender data = tfldata.napoleon_data1 template = napplot2;
   format order_subject subjfmt.;
run;
```

❶ The "RESPONSE_MARKERS" attribute map specifies the colors and the shape of the reason for treatment discontinuation, if applicable, or indicates ongoing. If a patient has a value of "Ongoing" or "O", then it means that the patient has not discontinued treatment. Even though the patient did not discontinue, a marker needs to be displayed to indicate that the patient is still on treatment. In the attribute map, there are two values or strings for each VALUE statement, such as "AE" and "Adverse Events", and "D" and "Death". The first value of each row matches the values in the NAPOLEON_DATA1 data set. In other words, the RESULT variable contains the values, "AE", "D", "O", and "PD". Therefore, when those values are encountered, the specified attributes are used to identify those values. The second values in the VALUE statement are used to create the values that appear on the legend. When specifying SAS colors, it is possible to use names such as GraphData4:color, or names of colors such as white, black, green, etc. It is also possible to use hexadecimal codes or one of the allowed seven color-naming schemes (Watson, 2020).

❷ The SCATTERPLOT statement plots the reason for treatment discontinuation markers. The markers are a square filled shape as specified by the "RESPONSE_MARKERS" attribute map, and they are 12pt in size.

❸ The DISCRETELEGEND statement displays the reason for treatment discontinuation legend on the bottom of the graph.

- The TYPE = MARKER option adds the markers to represent the treatment discontinuation reason or ongoing status. Because the treatment discontinuation markers are square, the markers representing the treatment discontinuation status in the legend are also square and look similar to the dose legend symbols in Output 4-1. If the treatment discontinuation status used circle markers as symbols, then the legend would have also displayed circular symbols.

- The EXCLUDE = ("AE" "D" "O" "PD") option excludes the entries relating to "AE", "D", "O", and "PD". As you recall, the "RESPONSE_MARKERS" attribute map had two values for each statement. Therefore, if the EXCLUDE option was not used, then the legend will have contained eight entries.

- The LOCATION = OUTSIDE option displays the legend outside of the wall area.

- The VALIGN = BOTTOM option displays the legend at the bottom of the graph (outside of the wall area).

- The DOWN = 1 option ensures all the treatment discontinuation statuses are displayed in one row.

- TITLE = "Reason for Treatment Discontinuation" adds the title above the four treatment discontinuation status values in the legend.

- By default, the legend title attributes are controlled by GraphLabelText, and the default size is 10pt. Using TITLEATTRS = (SIZE = 8PT) makes the legend title a little smaller than the default size specified in the style. For more information on styles, refer to Section 9.2.1 *Creating a Custom Style*.

- By default, the legend value attributes are controlled by GraphValueText, and the default size is 9pt. Using VALUEATTRS = (SIZE = 7PT) makes the legend values a little smaller than the default size specified in the style.

Output 4-2 includes the reason for treatment discontinuation and ongoing markers on top of the last stacked bar chart. In addition, Output 4-2 includes the treatment discontinuation status legend.

Output 4-2: Adding Reason for Treatment Discontinuation Markers and Legend

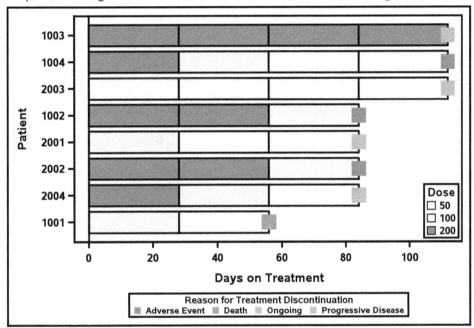

4.2.3 Adding Reason for Treatment Discontinuation Initial to the Plot

As indicated, each step builds upon the previous step. Program 4-3 will show you how to add to Program 4-2 and also display the reason for treatment discontinuation or ongoing initialism on top of the treatment discontinuation marker.

Program 4-3: Code for Adding Reason for Treatment Discontinuation Initial to the Plot

```
proc template;
    define statgraph napplot3;
        begingraph;

            << SAS CODE for attribute maps >>

            << SAS CODE for layout, bar chart and scatterplot statements >>

                scatterplot x = length y = order_subject /  ❶
                    markercharacter = result_label   ❷
                    markercharacterattrs = (size = 9pt);   ❸

            << SAS CODE for discrete legends >>

        endgraph;
    end;
run;

proc sgrender data = tfldata.napoleon_data1 template = napplot3;
    format order_subject subjfmt.;
run;
```

❶   The SCATTERPLOT statement plots the initialisms of the reason for treatment discontinuation or ongoing status.

❷   The MARKERCHARACTER option specifies that text from a variable should be displayed instead of using marker symbols. The RESULTS_LABEL variable contains the code letter(s) used to describe the treatment status, and therefore the letters are plotted instead of marker symbols.

❸   By default, the marker character attributes are controlled by GraphDataText, and the default size for the CustomSapphire style for GraphDataText is 7pt. Using VALUEATTRS = (SIZE = 9PT) makes the marker character attributes a little larger than they would have been by default.

In addition to Output 4-2, Output 4-3 includes the initialisms of the treatment discontinuation reason and ongoing status.

Output 4-3: Adding Reason for Treatment Discontinuation Initial to the Plot

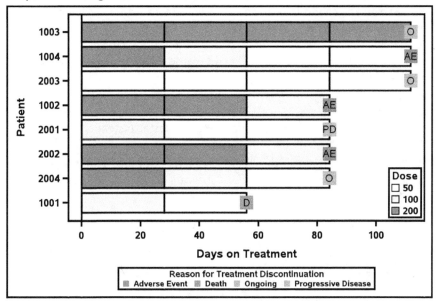

4.2.4 Adding Cohort information to the Plot

Program 4-4 will show you how to add to Program 4-3 and also display the cohort that each patient was in.

Program 4-4: Code for Adding Cohort information to the Plot (NAPOLEON_DATA1)

```
proc template;
   define statgraph napplot4;
      begingraph;

         << SAS CODE for attribute maps >>

         << SAS CODE for layout, bar chart and scatterplot statements >>

            scatterplot X = eval(length + 15) Y = order_subject / ❶
               markercharacter = cohort_label  ❷
               markercharacterattrs = (size = 9pt);  ❸

         << SAS CODE for discrete legends >>

      endgraph;
   end;
run;

proc sgrender data = tfldata.napoleon_data1 template = napplot4;
   format order_subject subjfmt.;
run;
```

❶ The SCATTERPLOT statement displays the patients' cohort information. The expression EVAL(LENGTH + 15) plots the cohort information at 15 units more than the value in the length column, on the X axis. Adding 15 creates a space between the bar chart and the cohort label. By default, the middle of the text is placed at the coordinates. Alternatively, using X = COHORT_LABEL_X would have also worked because the variable contains the same values as EVAL(LENGTH + 15).

❷ Again, the MARKERCHARACTER option specifies that text from a variable should be displayed instead of marker symbols. The COHORT_LABEL variable contains the patients' cohort information, and therefore the cohort information is plotted instead of marker symbols.

❸ Using VALUEATTRS = (SIZE = 9PT) makes the marker character attributes a little larger than they are specified by default. Therefore, the cohort labels are displayed a little larger than they would have been by default.

In addition to Output 4-3, Output 4-4 includes the patients' cohort information.

Output 4-4: Adding Cohort Information to the Plot

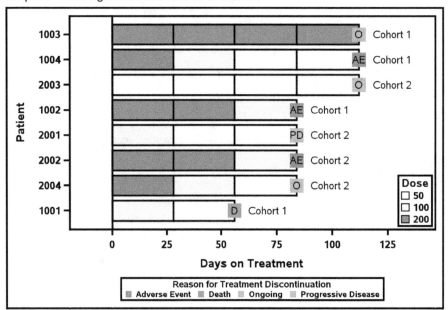

4.3 Creating a Napoleon Plot for Many Patients That Highlights Dose Interruptions

The Napoleon Plot in Output 4-4 is typically produced in the early phase studies at each data transfer. The patient numbers are relatively small, so it is quite easy to see the patient numbers on the graph. In later phases, such as Phases 3 and 4 studies, the patient numbers tend to be a lot higher, and therefore, you might need to come up with different strategies to create a clear Napoleon Plot, such as:

- Outputting the Napoleon Plots at a subgroup level, for example, a plot for males and a plot for females.

- Removing the outline from the bar charts in order to save space.

- Removing the patient numbers from the graph because SAS does not display them all. That is, SAS invokes the default option FITPOLICY = THIN when there is not enough room to fit all the tick values.

In later phases, there is the possibility that more patients will have dose delays. This section shows you how to create a Napoleon Plot to handle more patients, dose delays, and to add different annotations.

The data used for the Napoleon Plot is largely based on the EX transport file from the CDISC SDTM/ADaM Pilot Project. The EX transport file was used to create the ADEX data set, and the ADEX data set in conjunction with ADSL, helped to create the Napoleon data set, NAPOLEON_DATA2. The data set only contains the 84 patients that are in the xanomeline high dose treatment group. Display 4-2 shows a snippet of the data set. The variables in the data snippet are:

- SUBJID: the patient identifier.

- NUMBER: a variable that indicates the order the patients should be displayed on the graph. In Display 4-2, the data is sorted by the ascending values of the NUMBER variable so that patients who are on the study the longest are ordered last. This order is purposely done so that in the graph, the patients that have been on the study the longest are ordered first.

- SEX: the gender of the patient. "M" represents Male patients.

- VISITNUM: represents the visit number. You can see that patient 1106 had three distinct visits.

- ASTDY: represents the relative start day the patient started their dose. You can see that patient 1106 started their dose of 54 mg on day 1. The variable EXDOSE has the patient's dose information. Then patient's 1106 dose increased to 81 mg on day 17, and on day 175, patient 1288 had their dose reduced to 54 mg.

- AENDY: represents the relative end day that the patient ended their dose.

- EXDOSE: indicates the dose that the patient received.

- MAX_AENDY: represents the highest AENDY that each patient had. You can see that patient 1106 had the relative dose end date of 188 at visit number 12. 188 is the highest relative dose end day that the patient had, and therefore 188 is represented in the MAX_AENDY variable.

Patient 1288 in Display 4-2 shows an example of a dose delay. You can see that in the second record for patient 1288 (where VISITNUM is equal to 4), their relative end day was 156; however in the third record, patient 1288 has a relative start day of 176. Therefore, patient 1288 was off treatment for 19 days before they started their dose at their third visit. Even though patient 1288 had a dose delay for 19 days, the treatment duration for patient 1288 appears first on Output 4-5 because the order is based on duration on the study (the MAX_AENDY variable), and not the duration of treatment.

Display 4-2: Data Snippet Used for Napoleon Plots (NAPOLEON_DATA2)

Obs	SUBJID	NUMBER	SEX	VISITNUM	ASTDY	AENDY	EXDOSE	MAX_AENDY
1	1106	78	M	3	1	16	54	188
2	1106	78	M	4	17	174	81	188
3	1106	78	M	12	175	188	54	188
4	1408	80	M	3	1	20	54	189
5	1408	80	M	4	21	175	81	189
6	1408	80	M	12	176	189	54	189
7	1288	83	M	3	1	14	54	196
8	1288	83	M	4	15	156	81	196
9	1288	83	M	12	176	196	54	196

Program 4-5 shows you how to create a Napoleon Plot for a data set that has a large number of patients or when you want to display dose delays.

Program 4-5: Code for Napoleon Plot for Many Male Patients That Highlights Dose Interruptions

```
proc template;
    define statgraph nap5;
        begingraph;

            symbolimage name = completed  ❶
                image = "C:\Users\YEX7977\Documents\SAS Book\Chapter 4\
                        images\Kliponious-green-tick.png";  ❷
            discreteattrmap name = "Dose_Group";  ❸
                value "54" / fillattrs = (color = orange)
                            lineattrs = (color = orange pattern = solid);
                value "81" / fillattrs = (color = red)
                            lineattrs = (color = red pattern = solid);
            enddiscreteattrmap;
            discreteattrvar attrvar = id_dose_group var = exdose attrmap = "Dose_Group";
```

```
            legenditem type = marker name = "54_marker" / ❹
               markerattrs = (symbol = squarefilled color = orange)
               label = "Xan 54mg";

            legenditem type = marker name = "81_marker" / ❹
               markerattrs = (symbol = squarefilled color = red)
               label = "Xan 81mg";

           legenditem type = marker name = "completed_marker" / ❺
              markerattrs = (symbol = completed size = 12px)
              label = "Completed";

            layout overlay / yaxisopts = (type = discrete   ❻
                                          display = (line label)
                                          label = "Patient")
                             xaxisopts = (label = "Treatment duration
                                          from first dose date (Months)");
               highlowplot y = number ❼
                            high = eval(aendy/30.4375)
                            low = eval(astdy/30.4375) /
                    group = id_dose_group
                    type = bar ❼
                    lineattrs = graphoutlines ❼
                    barwidth = 0.2; ❼
                  scatterplot y = number x = eval((max_aendy + 10)/30.4375) / ❽
                     markerattrs = (symbol = completed size = 12px);
                  discretelegend "54_marker" "81_marker" "completed_marker" /  ❾
                     type = marker
                     autoalign = (bottomright) across = 1
                     location = inside title = "Dose";
            endlayout;
         endgraph;
      end;
   run;


proc sgrender data = tfldata.napoleon_data2 template = nap5; ❿
   format number subfmt.;  ❿
   where sex = "M";
   run;
```

❶ The SYMBOLIMAGE enables you to define a marker symbol using an image. The NAME = COMPLETED option specifies that the new marker symbol will have the identification equal to COMPLETED so that it can be referenced in other GTL statements.

❷ The IMAGE option in the SYMBOLIMAGE statement specifies the location and the name of the new marker. In this example, the image is a green checkmark.

❸ The DISCRETEATTRMAP creates the attributes of the doses the patients took in the xanomeline high dose group. Orange bar charts represent the period when the patients received the 54 mg dose, and red represents the 84 mg dose.

❹ The first two LEGENDITEM statements help to build the DISCRETELEGEND statement and hence the plot legend. The LEGENDITEM is used so that the markers that represent the dose and the markers that represent the patients that completed the study are in the same legend, and also so that the markers that represent the dose can be labeled as desired. If the legend was created from the fill colors that were used in the HIGHLOW statement, then the legend representing the doses would have displayed 54 and 81. However, displaying another legend label was desired. In the first LEGENDITEM statement, the TYPE = MARKER specifies that the legend displays a marker, and the NAME = "54_marker" option specifies that the legend can be identified by the name "54_marker". The attributes of the first LEGENDITEM statement have an orange color and the square shape. The label has the text "Xan 54mg" and "Xan 54mg" will be displayed on the legend.

❺   The SYMBOL = COMPLETED within the LEGENDITEM statement specifies that the legend will comprise of the newly created symbol, and the SIZE = 12px specifies that the symbol size will be larger than the default size.

❻   The TYPE = DISCRETE option specifies that the Y axis should be treated as discrete values instead of continuous or linear values. Since the values on the Y axis are numeric, by default SAS uses a linear axis; therefore, you want to override the default behavior. The discrete values are useful for the Y axis because the male patient identifiers are not sequential, so if the TYPE = LINEAR is used, the Y axis would have gaps. Because there are a lot of male patients, there is not enough room to display all the patient identifiers on the graph. Therefore, the DISPLAY = (LINE LABEL) option is used to only display the Y-axis line and label, and not display the patient identifiers. You can see that the REVERSE = TRUE option is not used and this is because in the data set, patients that have been on the study the longest are at the bottom of the data set.

❼   The HIGHLOW statement plots the dose duration for each patient. In the data set that is used for the plot, some patients have three rows for their dose duration. Having three rows is what allows the high-low plot to display a patient starting on 54 mg, then going up to 81 mg for the majority of the time, and then finally going back down to 54 mg dose. As mentioned, the NUMBER variable represents the patient identifier. AENDY represents the relative end day of the patient's dose and the HIGH option specifies the upper value to use in the HIGHLOW plot. HIGH = EVAL(AENDY/30.4375) indicates the upper value to use should be the relative end day / 30.4375. Dividing the relative day be 30.4375 enables you to see the treatment durations on a monthly interval. ASTDY represents the relative start day of the patient's dose and the LOW option specifies the lower value to use in the HIGHLOW plot.

- The TYPE = BAR option specifies that bar and line attributes represent the data.

- LINEATTRS = GRAPHOUTLINES specifies that the GRAPHOUTLINES element represents the line attributes. The default attributes for GRAPHOUTLINES in the CUSTOMSAPPHIRE style has a black line, and a line thickness of 2px. In section 9.2.1 *Creating a Custom Style*, Display 9-1 illustrates what is meant by GRAPHOUTLINES and you can see an example of modifying GRAPHOUTLINES in Section 9.2.1.3 *Modified CustomSapphire Style*.

- BARWIDTH = 0.2 specifies that the width of the bar is to be smaller than the default 0.85 bar width.

❽ The SCATTERPLOT statement plots the image of the green checkmarks that represent the patients who have completed the study. As previously mentioned, the MAX_AENDY variable represents the highest AENDY that each patient had. The expression EVAL((MAX_AENDY + 10)/30.4375) helps to position the green checkmark values a little higher than the high low bars on the X axis. Similarly to the LEGENDITEM option, the SYMBOL = COMPLETED option specifies that the marker will comprise of the newly created symbol – the green checkmark.

❾ The DISCRETELEGEND statement references the three items that were created using the LEGENDITEM statements, and creates a marker type legend in the bottom right-hand corner of the graph, inside the wall area, where the legend items are all in one column and having the title of "Dose".

❿ PROC SGRENDER creates the Napoleon Plot on the male patients. The FORMAT SUBFMT. statement potentially displays the patient numbers instead of the number that represents the patient order. However, because the tick values and tick have been suppressed in the Y-axis options, the patient numbers are not displayed.

Output 4-5 shows you the Napoleon Plot of male patients. You can see that there are many patients, so the patient labels have been suppressed. You also see that the patient that has been on the study the longest has a dose delay indicated by the white space between the red and orange bars.

Output 4-5: Napoleon Plot for Many Male Patients That Highlights Dose Interruptions

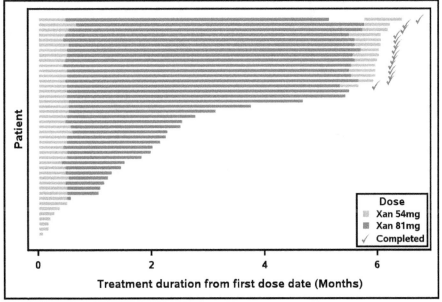

4.4 Chapter Summary

The Napoleon Plot is a handy plot for assessing treatment safety at each interim. For each patient, the extent of exposure, the reason for treatment discontinuation and the assigned cohort is displayed. Although, recall the true spirit of the Napoleon Plot stands for displaying several types of data conveniently on one single plot, and therefore anything that you deem as relevant can be displayed on the plot.

If you want to display the treatment durations for many patients, then consider subsetting the population in order to see the durations and the colors of the doses more easily.

A potential drawback of the BARCHARTPARM method used to display patients that had been dose reduced on the Napoleon Plot is that the treatment cycles have to be entered manually in the VALUE statement within the DISCRETEATTRMAP statement. Recall, in the examples, the strings in the VALUE statement for the attribute map that took into account the visit and dose looked similar to:

```
Value "1-50" "2-50" "3-50" "4-50" …;
```

Therefore, if there were too many treatment cycles in the study, which is common in long-term oncology studies, then each cycle would have to be entered manually in the VALUE statement. However, you can use a data-driven attribute map data set to circumvent this, and there are examples of this approach in Chapter 6, *Using SAS to Generate Graphs for Tumor Response.*

To be able to plot the Napoleon Plot, the data would need to be structured in a very certain way, as seen in Display 4-1. It is also a good idea to choose a color palette and or patterns that make it easy to distinguish a change in the patient's dose.

References

Harris, K. (2014). "Napoleon Plot for PharmaSUG." Proceedings of the Annual Pharmaceutical SAS Users Group Conference. Cary, NC: SAS Institute Inc. *PharmaSUG* (p. 1). San Diego: PharmaSUG.

Watson, R. (2020). "What's Your Favorite Color? Controlling the Appearance of a Graph." Proceedings of the *SAS Global Forum 2020 Conference.* Cary, NC: SAS Institute Inc.

Chapter 5: Using SAS to Generate Graphs for Time-to-Event Endpoints

5.1 Introduction to Time-to-Event Endpoints

In Chapter 2, you saw how you could use Kaplan-Meier curves to analyze the time to the first Dermatological Event. In this chapter, the time-to-event analysis will be explored more. You will see how you can create graphs that help you to analyze the common oncology endpoints: overall survival (OS) and progression-free survival (PFS). An overall survival example will be shown, and because the data for the overall survival and progression-free survival is in a similar format, it is easy for you to replicate the overall survival examples on progression-free survival data.

In Chapter 2, you saw how PROC LIFETEST could create Kaplan-Meier graphs. You also saw how to edit the template behind PROC LIFETEST so that you are able to customize your Kaplan-Meier graphs to meet your needs. In this chapter, you will learn how to create Kaplan-Meier graphs with SGPLOT and GTL. You will also learn how to create Forest Plots.

Generally, when analyzing time-to-event data, you would use the ADaM basic data structure (BDS) with the time-to-event components. A lot of the time, this data set has the name or prefix ADTTE. The data set used in this chapter is named ADTTEEFF. The ADTTEEFF data set is based on the CDISC pilot ADTTE data set. However, the survival times within the data set have been simulated.

5.1.1 Using ODS Output Objects for Time-to-Event Endpoints

When creating time-to-event figures, you would generally use your ADaM BDS with the time-to-event components, such as ADTTEEFF, with the LIFETEST and PHREG procedures. Each procedure used in conjunction with the appropriate ODS OUTPUT statements enable you to produce output objects, also known as output data sets. If you do not use the default graphs produced by PROC LIFETEST and PROC PHREG, then the output data sets contribute heavily to making your time-to-event figures.

While there are different output objects that can be produced from the different procedures, Table 5-1 shows the ODS Table Names for some that are commonly used for time-to-event figures.

Table 5-1: Common ODS Table Names for Time-to-Event Figures

Procedure	ODS Table Names	Description
LIFETEST	CensoredSummary	Contains the number of patients, event counts, censored counts, and the percentage of censored patients for each (treatment) group
LIFETEST	HomTests	Contains the log-rank p-value
LIFETEST	Quartiles	Contains the median survival times (of each treatment group)
LIFETEST	SurvivalPlot	Contains the survival probabilities over time (for each treatment). This is useful to plot the Kaplan-Meier curves
PHREG	HazardRatios	Contains the hazard ratios of specified comparisons

5.1.2 Understanding Overall Survival Endpoint

Firstly, it is important to know what overall survival (OS) and progression-free survival (PFS) means. According to the U.S. Department of Health and Human Services (2020), OS is defined as the time from randomization until death from any cause and is measured in the intent-to-treat population. Survival is considered the most reliable cancer endpoint, and when studies can be conducted to adequately assess survival, it is usually the preferred endpoint. This endpoint is precise and easy to measure since it is documented by the date of death or the censored date for patients who have not died by the end of the study. The censored date could be the last known alive date, study completion date, or any other relevant date specified by the sponsor.

Demonstration of a statistically significant improvement in overall survival can be clinically significant if the toxicity profile is acceptable and has often supported new drug approval. You can think of the toxicity profile as the safety profile, and it is generally acceptable when the number of adverse events and abnormal lab results in the treatment group are not deemed as being unsafe.

Although OS is probably the most reliable cancer endpoint, there are some difficulties in performing and analyzing survival studies, such as long follow-up periods in large trials and subsequent cancer therapy potentially confounding survival analysis.

5.1.3 Understanding Progression-Free Survival Endpoint

According to the U.S. Department of Health and Human Services (2020), PFS is defined as the time from randomization until objective tumor progression or death, whichever occurs first. The precise definition of tumor progression is detailed in each protocol. Sometimes the RECIST criteria are used to identify progression. PFS can reflect tumor growth and be assessed before the determination of a survival benefit.

Unlike OS, its determination is not confounded by subsequent therapy. For a given sample size, the magnitude of effect on PFS can be larger than the effect on OS. For example, the differences between the median survival times of patients on treatment compared to placebo can be larger when examining PFS compared to OS. Due to cancer trials often being small, and proven survival benefits of existing drugs are generally modest, treatment effects measured by PFS can be used as a surrogate endpoint. This can help to support accelerated approval, a surrogate endpoint to support traditional approval, or it can represent a direct clinical benefit.

5.1.4 Understanding the Usefulness of Kaplan-Meier Plots

It is important to know why Kaplan-Meier graphs are useful. Kaplan-Meier graphs are useful because they visually show the probability of an event at a given time interval. For OS, the Kaplan-Meier curve shows you the probability of survival. With this, you can analyze the efficacy of the treatments. As you have already read, having an efficacious treatment, as well as a treatment that meets the safety requirements, helps to get the treatment approved by regulatory agencies.

It is also interesting to know how the Kaplan-Meier (survival) function works, and you will see an example of this method in Display 5-2. However, first of all, you will see a data snippet of the OS data from the ADTTEEFF data set, as shown in Display 5-1. If you recall, there are 86 patients in the placebo group. So, at the start of the study there are 86 patients at-risk in the placebo group. Display 5-1 shows the first 11 patients in the placebo group, which either have an event (CNSR = 0) or are censored (CNSR = 1). You will notice that one patient (01-701-1015) has an event on Day 5. Five patients have an event on Day 6. Four patients have an event on Day 7, and two patients are censored on Day 7. It is possible to calculate the survival probability from the information in Display 5-1, and you will see how this is done in Display 5-2.

Display 5-1: Data Snippet of Overall Survival Data (ADTTEEFF)

Obs	USUBJID	PARAM	TRTP	TRTPN	ADT	AVAL	CNSR	EVNTDESC
1	01-701-1015	Overall Survival time	Placebo	0	03JAN2014	5.0	0	Death
2	01-703-1042	Overall Survival time	Placebo	0	31AUG2013	6.0	0	Death
3	01-705-1349	Overall Survival time	Placebo	0	08SEP2013	6.0	0	Death
4	01-708-1286	Overall Survival time	Placebo	0	12SEP2013	6.0	0	Death
5	01-718-1150	Overall Survival time	Placebo	0	15MAR2013	6.0	0	Death
6	01-704-1435	Overall Survival time	Placebo	0	12JAN2013	7.0	0	Death
7	01-705-1282	Overall Survival time	Placebo	0	24JUN2013	7.0	0	Death
8	01-708-1158	Overall Survival time	Placebo	0	22MAR2014	7.0	1	Alive
9	01-716-1024	Overall Survival time	Placebo	0	20JAN2013	7.0	0	Death
10	01-716-1108	Overall Survival time	Placebo	0	01MAY2013	7.0	1	Alive
11	01-718-1139	Overall Survival time	Placebo	0	17NOV2013	7.0	0	Death

Display 5-2 shows a data snippet of the survival probability for the patients on placebo. You can see that there are 86 patients participating in the study because at time point 0, at the start of the study, there are 86 patients "at-risk". You can also see that at time point 0 the survival probability is 1. The survival probability is 1 because no patients have had an event yet. On Day 5, one patient has an event, so the survival probability in the placebo treatment group is now 0.988. As seen on the Kaplan-Meier estimator page found on Wikipedia, the survival probability is calculated by using the previous survival probability when there was a death, if there is a previous survival probability available, multiplied by (1 – (number of deaths at that time point / patients known to have survived up until that time point)) (Wikipedia, 2020). In Display 5-2, the EVENT column represents the number of deaths, and the AT-RISK column represents the patients known to have survived. Patients known to have survived up until that time point are defined as patients that have not died or patients that have not been censored. Therefore, the survival probability on Day 5 is calculated as 1 – (1/86) = 0.988. If you recall in Display 5-1 patient (01-701-1015) had an event on Day 5.

On Day 6, four patients have an event, so the survival probability is now 0.988 * (1 – (4/85)) = 0.942. Please recall that 0.988 was the survival probability at Day 5, that is, the previous survival probability when there was a death.

Display 5-2: Data Snippet of Survival Probability for the Placebo Treatment Group

Time	Survival	At-Risk	Event	Censored
0	1	86	0	0
5	0.988	86	1	0
6	0.942	85	4	0
7	0.895	81	4	2
8	0871	75	4	2

5.2 Survival Curves for Time-to-Event Endpoints

Now that you know why Kaplan-Meier graphs are practical and how the survival function works, it is beneficial to know how to plot the graphs using PROC SGPLOT and GTL. With PROC SGPLOT you can create single-cell survival plots that include the patient's at-risk table and with GTL you can create single-cell and multi-cell survival plots.

5.2.1 Generating a Survival Plot with SGPLOT

You can create a Kaplan-Meier graph using PROC SGPLOT, as shown in the Survival Plot blog by Sanjay Matange (2014). Program 5-1 shows you how to do this as well.

Program 5-1: Creating a Survival Plot with SGPLOT

```
proc format;   ❶
value $trt
   "0" = "Placebo"
   "54" = "Low Dose"
   "81" = "High Dose";
run;

ods output survivalplot = survivalplot;   ❷
proc lifetest data = adam.adtteeff plots = survival (atrisk = 0 to 210 by 30);
   time aval * cnsr(1);
   strata trtpn;
run;

title1 'Product-Limit Survival Estimates';
title2 'With Number of Patients at-Risk';

proc sgplot data=SurvivalPlot;
   step x = time y = survival / group = stratum
                          name = 'survival';   ❸
   scatter x = time y = censored / markerattrs = (symbol = plus color = black)
                          name = 'censored';   ❹
   scatter x = time y = censored / markerattrs = (symbol = plus) group = stratum;❺


   xaxistable atrisk / x = tatrisk
                    class = stratum
                    location = inside
```

```
                        colorgroup = stratum
                        separator;   ❻
    keylegend 'censored' / location = inside position = topright;   ❼
    keylegend 'survival';   ❽

    yaxis min = 0;
    xaxis values = (0 to 210 by 30) label = "Days from Randomisation";
    format stratum $trt.;
run;
```

❶ PROC FORMAT maps the values "0", "54", and "81" to "Placebo", "Low Dose", and "High Dose" respectively so that these formatted values are eventually displayed on the graph.

❷ With the use of ODS objects, you can save the PROC LIFETEST outputs as a SAS data set, such as the SURVIVALPLOT table. This table contains the survival probabilities over time for each treatment. You can also use the SURVIVALPLOT table results to generate your own Kaplan-Meier graph, as shown in Program 5-1.

❸ The STEP statement produces a graph that "steps" down (or up) at each data point. The STEP statement within the SGPLOT procedure is used to produce the Kaplan-Meier Curve. The GROUP = STRATUM option allows for the Kaplan-Meier curve to be created for each treatment. The NAME = "SURVIVAL" option sets up a reference that is used later by the KEYLEGEND statement to produce the treatment legend.

❹ The use of the first SCATTER statement is solely to set up a reference, and in combination with the KEYLEGEND statement, produces the censored value legend. A black + symbol denotes the censored value. The COLOR = BLACK and the SYMBOL = PLUS options within the MARKERATTRS option helps to create the black + symbol. A black symbol is used because it is a neutral color and can represent the patients' censored values from any of the three treatment groups.

❺ The use of the second SCATTER statement along with the GROUP = STRATUM option is to produce the censored markers on the curve for each treatment. The censored markers match the color of the curve, which are blue, red, and green. Like the first SCATTER statement, the marker symbol is set to be a plus sign for all treatments. With the use of the MARKERATTRS = (SYMBOL = PLUS), this ensures that all the treatments have the same symbol rather than using the default symbol for grouped data.

❻ The XAXISTABLE statement produces the patients at-risk table. This table shows the number of patients still at-risk of having the event at chosen intervals. In the XAXISTABLE statement, you need to specify a variable that contains the values that you want to plot. The ATRISK variable contains all the patients' at-risk values, and therefore this variable is used to plot the at-risk values on the table.

- The X = TATRISK option plots the at-risk values along the X axis, as specified in the TATRISK column. The TATRISK variable contains the values from 0 to 210 in increments of 30, and therefore the at-risk values are displayed every 30 days from 0 to 210. The values in the TATRISK column have come from using the ATRISK = 0 to 210 by 30 option in PROC LIFETEST.

- The CLASS = STRATUM option creates a separate axis table for each of the three treatments specified in the STRATUM variable.

- The LOCATION = INSIDE option plots the axis table inside the plot area. If you want to plot the axis table outside the plot area, then you can use LOCATION = OUTSIDE.
- The COLORGROUP = STRATUM option colors the at-risk values differently for each of the treatments in the STRATUM variable. The COLORGROUP = STRATUM option is used so that the colors of the at-risk values could match the colors of the survival curves. Therefore, you can easily associate the treatments that relate to the survival curves and the at-risk values.
- The SEPARATOR option draws the line that separates the plots from the axis table when the axis table is inside the plot area.

❼ The first KEYLEGEND statement along with the LOCATION = INSIDE and POSITION = TOPRIGHT options plots the censored symbol from the first SCATTER statement in the top right section of the graph.

❽ The second KEYLEGEND statement produces the treatment legend. The LOCATION and POSITION options are not specified, and therefore, the default options are used. These default options make the legend appear outside of the main plot area and at the bottom.

Output 5-1 shows the Kaplan-Meier plot of overall survival stratified by the treatment. The plot shows that patients who had the high dose survived the longest, followed by patients who had low dose, and then followed by the placebo patients.

Output 5-1: SGPLOT Survival Plot

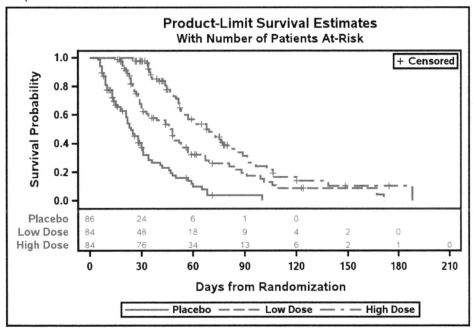

5.2.2 Generating Single-Cell Survival Plot with Graph Template Language

Frequently, customers request to see the log-rank *p*-value, hazard ratio, median survival time and other statistics alongside the Kaplan-Meier graph. Adding these statistics to the survival plot are easily done with the use of GTL. Although first, you would need to calculate the necessary statistics. You can also use PROC SGPLOT and the INSET statement to insert text into graphs, and if you only want to add one statistic or one block of text, such as "Median = XX.XX", then using PROC SGPLOT is simpler. Also, if the text you wanted to display is in a variable, then you can use the PROC SGPLOT and the TEXT statement to plot the text on the graph. However, if the text that you want to add is not a variable in the data set, or you want to add more text, or you want to align the text a certain way, then using GTL is the better option because of its flexibility.

When you are adding statistics such as the log-rank *p*-value and hazard ratio onto the graphs, a good approach is to create macro variables for storing the statistics, and then using the ENTRY statement or TEXTPLOT statements to display the statistics. This approach is good because these statements are quite flexible and enable you to place your text on your graph wherever you want.

As previously mentioned, the hazard ratios and log-rank *p*-values are frequently requested alongside Kaplan-Meier graphs. You can use the PHREG procedure to obtain the hazard ratios and the LIFETEST procedure to calculate the log-rank *p*-value and obtain the data for the Kaplan-Meier curves. Program 5-2 shows how you can add the log-rank *p*-value and hazard ratios onto your Kaplan-Meier plot.

Program 5-2: Creating Survival Plot with GTL

```
<< SAS CODE for creating treatment format >>   ❿

ods output SurvivalPlot = SurvivalPlot;
ods output HomTests = HomTests(where = ( test = "Log-Rank"));   ❶
proc lifetest data = adam.adtteeff plots=survival(atrisk = 0 to 210 by 30);
   time aval * cnsr(1);
   strata trtpn;
run;

ods output HazardRatios = HazardRatios;   ❷
proc phreg data = adam.adtteeff;
   class trtpn /ref=last;
   model aval * cnsr(1) = trtpn;
   hazardratio trtpn /diff = ref;
   format trtpn trt.;
run;

proc sql;
   select put(ProbChiSq, PVALUE6.4) into: log_rank_pvalue   ❸
   from HomTests;
   select put(HazardRatio, 6.2) into: HazardRatio1-:HazardRatio2   ❹
   from HazardRatios;
quit;

proc template;
   define statgraph kmtemplate;
      mvar log_rank_pvalue HazardRatio1 HazardRatio2;   ❺
      begingraph;
```

```
        entrytitle "Product-Limit Survival Estimates" /
            textattrs=(size=11pt);
        entrytitle "With Number of Patients at-Risk" / textattrs=(size=10pt);
          layout overlay / xaxisopts = (label = "Days from Randomisation"
            linearopts = (tickvaluesequence = (start = 0 end = 210
                                                increment = 30)));

              stepplot x = time y = survival / group = stratum
                                      name = "Survival"
                                      legendlabel = "Survival";   ❻

              scatterplot x = time y = censored / markerattrs = (symbol = plus
                                                        color = black)
                                      name = "Censored";   ❼
               scatterplot x = time y = censored /
                                      markerattrs = (symbol = plus)
                                      group = stratum;   ❼

              discretelegend "Censored" / location = inside
                                      autoalign = (topright);
              discretelegend "Survival" / location = outside
                                      across = 3 down = 1;

              InnerMargin / align = bottom opaque = true Separator = true;
                 AxisTable Value = AtRisk X = tAtRisk / class = stratum
                                                colorgroup = stratum;
              EndInnerMargin;

              layout gridded / columns = 2 rows = 3 border = true
                             halign = right valign = top
                             outerpad = (top = 25px);   ❽
                 entry halign = right "Log-Rank: "
                    textattrs = (style = italic) "p"
                    textattrs = (style = normal) "-value = ";   ❾
                 entry halign = left log_rank_pvalue;
                  entry "HR: High Dose vs Placebo = ";
                  entry halign = left HazardRatio1;
                  entry "HR: Low Dose vs Placebo = ";
                  entry halign = left HazardRatio2;
                endlayout;

          endlayout;
        endgraph;
      end;
  run;

  proc sgrender data = Survivalplot template = kmtemplate;   ❿
     format stratum $trt.;
  run;
```

❶ As illustrated in Program 5-1, the survival probabilities can be captured in a SAS data set using the ODS statement. In Program 5-2, PROC LIFETEST outputs the HOMTEST table, which contains the log-rank *p*-value.

❷ PROC PHREG calculates and outputs the hazard ratios in the HAZARDRATIOS table. The HAZARDRATIOS table contains the hazard ratios of the high and low dose against the placebo.

- The CLASS statement identifies the qualitative variables that you want to use in the analysis. TRTPN represents the numeric values of the three treatments, and this is specified as the qualitative variable so that hazard ratios can be calculated between the treatments. TRTPN is used in the CLASS statement instead of TRTP because TRTPN sorted the treatments in the necessary order, which helps obtain the hazard ratios in the correct sequence. The REF = LAST option appoints the last ordered level as a reference and therefore appoints the value 3 (the placebo group) as a reference. The appointment of this reference helps when calculating the hazard ratios.

- In PROC PHREG, two forms of model syntax can be specified. The syntax for the MODEL in this example is as follows: **MODEL response <*censor (list)> = effects </ options>.** This syntax is used because there is only one response variable in this analysis, which is the overall survival times of the patients. The overall survival times are located in the AVAL variable. The response variable is required to contain nonnegative values. The censor variable is optional; however, if it is present, it should be numeric. In Program 5-2, the censoring variable is called CNSR, and the censored values are denoted by 1. Therefore, the following code is used to represent the censor list portion of the model: CNSR(1). TRTPN is specified as the effect variable so that the hazard ratios can be calculated between the treatments.

- The HAZARDRATIO statement enables you to request hazard ratios on the effect variables in your model. TRTPN is specified so that hazard ratios can be calculated between the treatments, and the DIFF = REF option requests comparisons between the reference level and all other levels of the CLASS variable, and therefore in this example comparisons are made between treatments 1 and 3, and between treatments 2 and 3.

- The FORMAT TRTPN TRT. is used so that the descriptions of the hazard ratios display the treatment name instead of the numerical values.

❸ PROC SQL uses the INTO: clause to create the *LOG_RANK_PVALUE* macro variable, which contains the value of the log-rank *p*-value.

❹ The two hazard ratio values are stored as macro variable values. There are two hazard ratio values because one of the hazard ratios are from the comparison of high dose against the placebo, and the other hazard ratio is from the low dose against the placebo comparison.

❺ The MVAR statement defines three macro variables in the KMTEMPLATE template. The first macro variable is *LOG_RANK_PVALUE*. *LOG_RANK_PVALUE* contains the log-rank *p*-value and is <.0001 in Output 5-2. The second and third macro variables have the hazard ratio values, which are 0.20 for high dose versus placebo and 0.36 for low dose versus placebo in Output 5-2.

❻ The STEPPLOT statement produces the Kaplan-Meier Curve. The NAME = "Survival" option sets up a reference, which is used later by the DISCRETLEGEND statement to produce the treatment legend. This statement is equivalent to the STEP statement in Program 5-1.

❼ These SCATTERPLOT statements produce the censored value legend, and the censored markers on the curve for each treatment. These statements are equivalent to the SCATTER statements in Program 5-1.

❽   The LAYOUT GRIDDED statement sets up a container that has two columns and three rows and is aligned in the top right-hand corner of the graph. The ability to specify rows and columns helps the presentation of the text (or graph) within the container.

❾   The ENTRY statements display the log-rank *p*-value and hazard ratios on the graph. With the ENTRY statements, there are options available to horizontally align the text in the left, center or right. In this example, the HALIGN = RIGHT prefix option aligns the "log-rank" text and the subsequent text to the right. There are options available to italicize the text, amongst other options. In this example, TEXTATTRS = (STYLE = ITALIC) "p" italicizes the letter "p".

❿   A character format is created because PROC LIFETEST creates a STRATUM variable with character versions of the values that are in TRTPN, such as "0", "54", and "81". To display the formatted values instead of "0", "54" and "81", the FORMAT STRATUM $TRT. statement is used. The code to create the format can be found at https://github.com/rwatson724/SAS-Graphs-Clinical-Trials-Example.

Output 5-2 shows the Kaplan-Meier curve, log-rank *p*-value and hazard ratio results. The log-rank *p*-value of <.0001 indicates that, if the alpha level relating to the log-rank test is 5%, then statistically there is a difference in the survival curves between the three treatments, and this is because the *p*-value is less than 5%. The alpha level of 5% is generally used, although other alpha levels such as 10% or 1% could also be specified in the statistical analysis plan or by the statistician.

Output 5-2: GTL Survival Plot

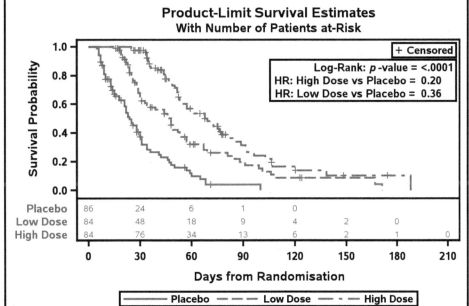

In the article titled *Survival plots of time-to-event outcomes in clinical trials: good practice and pitfalls* (Pocock, Clayton and Altman, 2002), it is recommended that instead of producing survival plots "stepping" down, that the plots should "step" up. A plot "stepping" up shows the cumulative proportion of patients experiencing the event by time. In principle, both contain the same information, but the visual perceptions about the comparison of treatment groups can be quite different, especially when the scale on the Y axis is reduced so that the Y axis only goes from 0 to the maximum cumulative incidence.

Since the crude cumulative incidence of an event at a given time is one minus the overall survival probability at that time, the crude cumulative incidence can be worked out and plotted quite easily. You could create a new variable in the SURVIVALPLOT data set, such as CUMULATIVE = 1 – SURVIVAL. Then in your GTL code, you could use the CUMULATIVE variable instead of the SURVIVAL variable. Alternatively, in GTL, you could use the EVAL function to calculate the crude cumulative incidence. Program 5-3 shows how you can use the EVAL function, and modify the code in Program 5-2, to create the crude cumulative incidence plot.

Program 5-3: Creating Cumulative Incidence Plot with GTL

```
proc template;
   define statgraph kmtemplate;
      mvar log_rank_pvalue HazardRatio1 HazardRatio2;
      begingraph;
         layout overlay / xaxisopts = (label = "Days from Randomisation"
                    linearopts = (tickvaluesequence = (start = 0 end = 210
                                                       increment = 30)))
                    yaxisopts = (label = "Crude Cumulative Incidence")   ❶;
               stepplot x = time y = eval(1-survival)   ❷/ group = stratum
                                                      name = "Survival";

            scatterplot x = time y = eval(1-censored)   ❸/
                  markerattrs = (symbol = plus color = black)
                              name = "Censored"
                  legendlabel = "Censored";   ❹
             scatterplot x = time y = eval(1-censored)   ❸/
                  markerattrs = (symbol = plus) group = stratum;

            discretelegend "Censored" / location = inside
                                      autoalign = (bottomright);
            discretelegend "Survival" / location = outside
                                      across = 3 down = 1;

            innermargin / align = bottom opaque = true Separator = true;
               axistable Value = AtRisk X = tAtRisk / class = stratum
                                                    colorgroup = stratum;
            endinnermargin;

            layout gridded / columns = 2 rows = 3 border = true
                            halign = right valign = bottom
                            outerpad = (bottom = 25px);
               entry halign = right "Log-Rank: "
                  textattrs = (style = italic) "p"
                  textattrs = (style = normal) "-value = ";
               entry halign = left log_rank_pvalue;

                entry "HR: High Dose vs Placebo = ";
                entry halign = left HazardRatio1;
```

```
                      entry "HR: Low Dose vs Placebo = ";
                      entry halign = left HazardRatio2;
                endlayout;

            endlayout;
        endgraph;
    end;
run;

proc sgrender data = Survivalplot template = kmtemplate;
    format stratum $trt.;
run;
```

❶ The LABEL option specifies that the Y axis has the label "Crude Cumulative Incidence". The label has been updated to reflect that the crude cumulative incidence, instead of the survival probabilities, will now be used in the graph.

❷ In Program 5-2, the STEPPLOT statement produced a "step-down" graph that used the values from the SURVIVAL variable. For this example, the desired graph is to produce a "step-up". In order to achieve this without having to do additional data processing in a separate DATA step, the EVAL function can be used to define that the values on the Y axis will be 1 – SURVIVAL.

❸ In order to coincide with the cumulative incidence, the censor markers need to be adjusted as well. The EVAL function is defining that markers on the KM plot will be 1 – SURVIVAL.

❹ You can use the LEGENDLABEL statement to label your legend when your legend is not based on a grouping variable. LEGENDLABEL = "Censored" labels the associated legend as "Censored". SCATTERPLOT statements that do not specify the group option and a LEGENDLABEL, and which use a variable as the Y axis, have a legend label based on the Y variable label if present, otherwise it would be based on the Y variable name. SCATTERPLOT statements that do not specify the group option and a LEGENDLABEL, and which use an expression as the Y axis, have a legend based on the expression. For example, if you use the expression EVAL(1 – CENSORED), then the legend would display "1 – CENSORED", unless you use the LEGENDLABEL option to specify a different legend. EVAL(1 – CENSORED) is used because the censoring values had to be adjusted to account for the "step-up" graph. With the "step-down" graph, a legend label did not need to be specified since the variable name, CENSORED, could be used.

Output 5-3 shows the Kaplan-Meier crude cumulative incidence curves, log-rank *p*-value and hazard ratio results. This output is similar to Output 5-2, except for the curves are "stepping" up instead of down on the Y axis.

Output 5-3: GTL Cumulative Incidence Plot

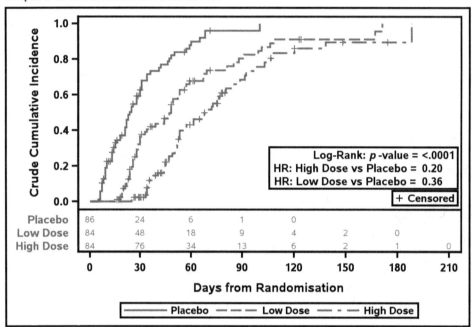

5.2.3 Generating Multi-Cell Survival Plots with Graph Template Language

In Section 5.2.2 *Generating Single-Cell Survival Plot with Graph Template Language*, you saw how to create single-cell survival plots with GTL. In this section, you will see how to develop multi-cell survival plots. The multi-cell survival plots enable you to create more than one survival curve on the same graph, and this is helpful for showing the results of the subgroup analysis or showing more than one endpoint on the same graph, such as overall survival and progression-free survival on the same graph. Using multi-cell plots also enable you to display the patients' at-risk table, below the X axis, which is a common request.

The median survival time is one way to see how well a new treatment works. In the context of overall survival, the median survival is the length of time from the randomization date that half of the patients are still alive.

5.2.3.1 Generating Survival Plot with Median Survival Next to the Survival Curve

The median survival time for each treatment can be displayed on the graph similarly to how the log-rank *p*-value and hazard ratios are displayed in Outputs 5-2 and 5-3. You can also visually highlight the median survival time on the plot. Program 5-4 shows you how to represent the median survival times with the DROPLINE statement, how you can use the LATTICE layout statement to create a multi-cell survival plot, and how to position the at-risk table outside of the plot area.

Program 5-4 is similar to Program 5-2. For example, as done in Program 5-2, the survival probabilities, *p*-values and hazard ratios are obtained using PROC LIFETEST and PROC PHREG and stored in macro variables for use in PROC TEMPLATE when defining the graph template that will be used to render the graph. Program 5-4 illustrates the additional components need to produce the median survival times. The complete program can be viewed at https://github.com/rwatson724/SAS-Graphs-Clinical-Trials-Example.

Program 5-4: Creating GTL Survival Plot with Median Survival Times

```
ods output Quartiles = Quartiles;      ❶
ods output SurvivalPlot = SurvivalPlot;
ods output HomTests = HomTests(where = (test = "Log-Rank"));
proc lifetest data = adam.adtteeff
              plots = survival(atrisk = 0 to 210 by 30);
   time aval * cnsr(1);
   strata trtpn;
run;

<< SAS CODE for PROC PHREG, and for creating macro variables>>

proc sql;
   select estimate,
          put(estimate, 4.1) into :MedianSurvival1-:MedianSurvival3   ❷,
                                  :CMedianSurvival1-:CMedianSurvival3   ❸
   from quartiles
   where percent = 50;
quit;

%macro km;   ❺

proc template;
   define statgraph kmtemplate;
      nmvar MedianSurvival1 MedianSurvival2 MedianSurvival3;   ❷
      mvar log_rank_pvalue HazardRatio1 HazardRatio2 CMedianSurvival1
         CMedianSurvival2 CMedianSurvival3;   ❸
      begingraph;
         layout lattice / rows = 2 rowweights = preferred;   ❹
            layout overlay / xaxisopts = (
                   label = "Days from Randomisation"
                   linearopts = (tickvaluesequence = (start = 0 end = 210
                                                      increment = 30)));

            %do i = 3 %to 1 %by -1;   ❺
               dropline y = 0.50 x = MedianSurvival&I   ❻   /
                   dropto = both   ❼
                   lineattrs = (thickness = 1px
                                color = graphdata&i:color   ❽
                   pattern = graphdata&i:linestyle)
                   label = CMedianSurvival&I;   ❾
            %end;
```

```
            << SAS CODE for KM Curve and legends >>


            << SAS CODE for log-rank and hazard ratio table >>

        endlayout;

        layout overlay / xaxisopts = (display = none
            linearopts =
                (tickvaluesequence = (start = 0 end = 210
                                      increment = 30)));   ❿
            axistable value = atrisk x = tatrisk / class = stratum
                                                colorgroup = stratum;

        endlayout;

      endlayout;
    endgraph;
  end;
run;

%mend;   ❺
%km;

proc sgrender data = Survivalplot template = kmtemplate;
    format stratum $trt.;
run;
```

❶   Using the ODS statement PROC LIFETEST outputs the QUARTILES table. This table contains the median survival times of each treatment group.

❷   The NMVAR statement is used to obtain the three median values from the macro variables *MEDIANSURVIVAL1*, *MEDIANSURVIVAL2*, and *MEDIANSURVIVAL3* and store them as a numeric value so that those values can eventually be used in the DROPLINE statement.

❸   The MVAR statement obtains the macro variable values from the macro variables *CMEDIANSURVIVAL1*, *CMEDIANSURVIVAL2*, and *CMEDIANSURVIVAL3* and stores them as a character value. This is so the median values are displayed to one decimal place on the plot beside the respective survival curves. The MVAR statement also obtains the values of the macro variables *LOG_RANK_PVALUE*, *HAZARDRATIO1*, and *HAZARDRATIO2* and stores them as character values. The difference between the NMVAR and the MVAR statements are that the NMVAR statement stores the values as numeric values and MVAR statement stores the values as character values. Some of the arguments expect a numerical result, such as the X-axis value in the DROPLINE statement. Therefore, if they expect a numerical result and you want to use macro variable values, then use the NMVAR statement. A consequence of using NMVAR when you want to use character values on your plot, is that the format of the values might not display how you intended them to display. For example, the value 48 might be shown instead of 48.0, although you expected the values to be displayed to 1 decimal place.

❹   One way to produce multi-cell charts is to use the LAYOUT LATTICE statement. The ROWS = 2 option creates two rows of cells where you are able to output graphs. In this example, the first row contains the survival curves, and the second row contains the at-risk numbers. The ROWWEIGHTS = PREFERRED option specifies that each row gets its preferred height, and this is automatically done for you. The other options are to specify the heights that you want each row individually or to use the ROWWEIGHTS = UNIFORM option to equally divides the total available height among all of the rows.

❺ The iterative DO statement helps to draw the lines highlighting the median for each treatment. If the iterative DO statement were not used, then there would have to be three separate DROPLINE statements to produce the lines, which highlight the median. The iterative DO statement starts at the highest value, which is three in this example, and goes down to one to ensure that the appropriate colors are shown horizontally, from the median survival probability to the three Kaplan-Meier curves. In order for the iterative DO statement to work, the template needs to be within a macro, and for that reason the **KM** macro is used.

❻ The DROPLINE statement has two required arguments. The arguments are the values of the X and Y coordinates. The Y value is 0.5 because that represents the median, and the value of X is the median survival time of each treatment group. The macro variable *MEDIANSURVIVAL&I* contains the median survival times for each of the three treatments.

❼ The DROPTO = BOTH option draws a line to both axes from the data points mentioned in callout ❻. If you only want do draw the line to one of the axis, then you can specify DROPTO = X or DROPTO = Y.

❽ The text GRAPHDATA&I:COLOR resolves to GRAPHDATA1:COLOR, GRAPHDATA2:COLOR, and GRAPHDATA3:COLOR based on the value of the macro variable *I*. The COLOR = GRAPHDATA&I:COLOR option within the LINEATTRS option helps to plot the lines the same color as the curves. This is because the curves also use the attributes GRAPHDATA1:COLOR, GRAPHDATA2:COLOR, and GRAPHDATA3:COLOR. Chapter 9 explains these style options further.

❾ The LINEATTRS option PATTERN = GRAPHDATA&I:LINESTYLE matches the patterns from the DROPLINE statement with the patterns on the survival curve for each treatment group. The LABEL = *CMEDIANSURVIVAL&I* option, displays the value of the median survival point on the plot, at the DROPLINE coordinates.

❿ The purpose of this LAYOUT OVERLAY statement is to produce the at-risk table. It is a good idea to keep the same X axis LINEAROPTS as was used in the first LAYOUT OVERLAY statement, that is, LINEAROPTS = (TICKVALUESEQUENCE = (START = 0 END = 210 INCREMENT = 30)), because this ensures that the at-risk numbers are placed at the correct time points. If you do not use the same LINEAROPTS, then there is a risk that the patients' at-risk values do not match up with the time on the X axis.

Output 5-4 is similar to Output 5-2, except that the plot displays the median survival time value for each treatment, the plot visually depicts the median survival time, and the plot displays the at-risk value outside of the plot area instead of inside. The visual depiction of the median survival time is apparent because a horizontal line is drawn from the Y axis where the survival probability is 50% until the line hits the Kaplan-Meier curve. Then once the Kaplan-Meier curve is hit, a vertical line is drawn to the X axis. You can see that the median survival times are 24, 48, and 69.1 days for the placebo, low dose, and high dose treatment group respectively. Patients on the high dose live for longer compared to the other treatments.

Output 5-4: GTL Survival Plot with Median Survival Times

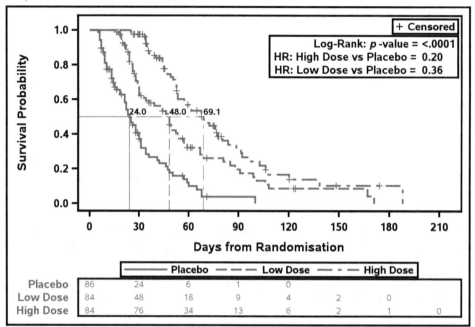

5.2.3.2 Generating Survival Plot with Median Survival and Event Statistics Table

There might be occasions when clients ask you to add a table of the median survival time and the number of events that each treatment had onto the Kaplan-Meier plot. In Chapter 2, you saw how you could use the *%SurvivalSummaryTable* macro along with the *%ProvideSurvivalMacros* and *%CompileSurvivalTemplates* macros to produce the summary table under the KM graph quickly. In this section, you will see how to use PROC LIFETEST to obtain the events from the ODS table CENSOREDSUMMARY, and how to display the median survival time and the events in a table. Some advantages of this method, compared to using the macros in Chapter 2, are that you will have more control over where to put the table and the values that you want to display in the table. For example, you can display the percentage of events, and label the column headers how you desire.

Program 5-4 showed you how to display the median survival time right beside the respective curve. However, on occasions, the different curves can be very close together, and therefore, it could be challenging to put the values next to the curves, because the values could overlap with each other. Therefore, in situations where the different curves are close to each other, it could be more beneficial to place the median survival values within a table. Program 5-5 shows you how you can create a survival plot that also show a table of median survival times and other statistics. You can view the complete Program 5-5 at https://github.com/rwatson724/SAS-Graphs-Clinical-Trials-Example.

Program 5-5: Creating GTL Survival Plot with Table of Median Survival Times

```
ods output Quartiles = Quartiles;
ods output SurvivalPlot = SurvivalPlot;
ods output CensoredSummary = CensoredSummary;   ❶
proc lifetest data = adam.adtteeff plots=survival(atrisk=0 to 210 by 30);
    time aval * cnsr(1);
    strata trtpn;
run;

proc sql;  ❷
    select put(total, 3.),
            put(failed, 3.)||" ("||put((100-pctcens), nlnum5.2)||")"
        into  :total1 - :total3,
                :evnt_perc1 - :evnt_perc3
    from CensoredSummary;
quit;

proc sql;  ❸
    select put(estimate, 4.1) into: CMedianSurvival1-:CMedianSurvival3
    from quartiles
    where percent = 50;
quit;

proc template;
    define statgraph kmtemplate;
        mvar total1 total2 total3 evnt_perc1 evnt_perc2 evnt_perc3  ❷
            CMedianSurvival1 CMedianSurvival2 CMedianSurvival3  ❸;
        begingraph;
            layout lattice / rows = 2 rowweights = preferred;
                layout overlay / xaxisopts = (label = "Days from Randomisation"
                linearopts = (tickvaluesequence = (start = 0 end = 210
                                                    increment = 30)));

                    << SAS CODE for KM Curve and legends >>

                    layout gridded / columns = 4 rows = 4 border = true  ❹
                                     autoalign = (topright)
                                     outerpad = (top = 25px);  ❹

                        entry halign = center "Planned Treatment" /  ❺
                                textattrs = (weight = bold);
                        entry halign = center "Patients" /  ❺
                                textattrs = (weight = bold);
                        entry halign = center "Events (%)" /  ❺
                                textattrs = (weight = bold);
                        entry halign = center "Median Time" /  ❺
                                textattrs = (weight = bold);

                        entry halign = center "Placebo" /  ❻
                           textattrs=(color=graphdata1:contrastcolor) ;
                        entry halign = center total1 /  ❻
                           textattrs=(color=graphdata1:contrastcolor);
                        entry halign = center evnt_perc1 /  ❻
                           textattrs=(color=graphdata1:contrastcolor);
                        entry halign = center CMedianSurvival1 /  ❻
                           textattrs=(color=graphdata1:contrastcolor);

                    << Repeated for low dose and high dose group >>
```

```
            endlayout;
          endlayout;

          << SAS CODE for patient at-risk table >>

        endlayout;
      endgraph;
    end;
run;

proc sgrender data = Survivalplot template = kmtemplate;
    format stratum $trt.;
run;
```

❶ Using the ODS statement, PROC LIFETEST outputs the CENSOREDSUMMARY table. This table contains the number of patients, event counts, censored counts, and the percentage of censored patients for each treatment group.

❷ PROC SQL creates macro variables (*TOTAL1, TOTAL2,* and *TOTAL3*), which contain the total number of patients in each treatment group. In addition, PROC SQL concatenates the event counts with the percentage of patients with events, for each treatment and stores the values in the macro variables *EVNT_PERC1, EVNT_PERC2,* and *EVNT_PERC3*. The concatenated values of the macro variables are in the format XXX (XX.XX), where XXX is the number of patients in the treatment group, and (XX.XX) is the percentage of patients with events in the treatment group. The MVAR statement in PROC TEMPLATE obtains the macro variable values from the macro variables *TOTAL1, TOTAL2, TOTAL3, EVNT_PERC1, EVNT_PERC2,* and *EVNT_PERC3* and stores them as character values. The MVAR statement stores the values so that they can be referenced by the ENTRY statements (shown in the full program on GitHub) and displayed within a table in the Kaplan-Meier plot.

❸ Similarly, to Program 5-4, the MVAR statement obtains the macro variable values from the macro variables *CMEDIANSURVIVAL1, CMEDIANSURVIVAL2,* and *CMEDIANSURVIVAL3* and stores them as a character value. The macro variable values are stored as character values, so that the median values are displayed to one decimal place as desired, and the MVAR statement is used so that the values can be eventually referenced within the program by the ENTRY statement (shown in the full program on GitHub) and therefore be displayed within a table beside the plot.

❹ The LAYOUT GRIDDED statement sets up a container that has four columns and four rows and is aligned in the top right- hand corner of the graph minus the padding. The OUTERPAD = (TOP = 25px) option adds a padding of 25px to the top of the layout. This top padding effectively pushes the layout down 25px, and leaves room for the Censored legend, which is positioned in the top right-hand corner of the graph.

❺ The first four ENTRY statements create the column headers for the tables. For the purpose of this output, they are referred to as column headers because they are created in the first row of the GRIDDED layout. The TEXTATTRS = (WEIGHT = BOLD) option bolds the text, to make the column headers stand out more.

❻ The remaining ENTRY statements display the macro variable values relating to the number of patients, the number of events, the percentage of the events, and the median survival time for each treatment. The TEXTATTRS = (COLOR = GRAPHDATA1:CONTRASTCOLOR) option makes the font color of the placebo statistics in the table, the same color as the placebo Kaplan-Meier curve and the placebo at-risk numbers. GRAPHDATA2:CONTRASTCOLOR and GRAPHDATA3:CONTRASTCOLOR are the colors that the low dose and high dose treatment groups are assigned to respectively.

Output 5-5 is similar to Output 5-4, except that the median survival times are shown in a table as opposed to directly beside the curve. Output 5-5 is also similar to Output 5-2 except that instead of showing a table of the log-rank *p*-values and the hazard ratios, a table of the number of patients, events, percentage of events, and median survival time for each treatment group is displayed.

Output 5-5: GTL Survival Plot with Table of Median Survival Times

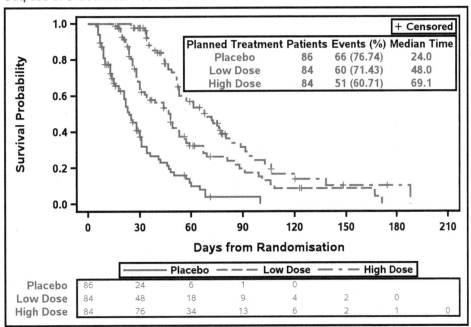

5.2.3.3 Generating Survival Plots for Subgroup Analysis

Subgroup analysis is often done on overall survival and progression-free survival to support product approval. An example of subgroup analysis is producing a Kaplan-Meier plot for each gender, that is, a Kaplan-Meier plot of female patients and another Kaplan-Meier plot of male patients. Typically, subgroup analysis is done to show that the treatment differences that have occurred in the overall population are consistent across the subgroups. In other words, the differences are not being skewed due to a subgroup having a more efficacious outcome. Sometimes the subgroup analysis is done to target a particular subgroup. For example, companies might want to verify that a treatment only works on males but not females.

According to Tanniou, van der Tweel, Teerenstra, and Roes (2016), there are four purposes for subgroup analysis. These are:

- Investigate the consistency of treatment effects across subgroups of clinical importance.

- Explore the treatment effect across different subgroups within an overall non-significant trial.

- Evaluate safety profiles limited to one or a few subgroup(s).

- Establish efficacy in the targeted subgroup when included in a confirmatory testing strategy of a single trial.

Performing subgroup analysis on overall survival and progression-free survival generally means producing Kaplan-Meier plots and survival estimates for each subgroup of interest. There are usually many subgroups in a study, so it could be efficient to use a macro when doing the subgroup analysis.

You can also save on the number of outputs produced by outputting a figure that has two or three subgroups on it. The LAYOUT LATTICE, LAYOUT DATAPANEL, and LAYOUT GRIDDED can produce graphs that show two or three subgroups, because these layouts can produce multi-cell graphs. When incorporating text in various areas of the graph, using the LAYOUT LATTICE statement to plot the survival analysis for each gender is more flexible than using the DATAPANEL and GRIDDED layout. This is because with the LAYOUT LATTICE you can use different macro variables for the different subgroups, and therefore you can display the statistics such as the log-rank and hazard ratios for each subgroup. Program 5-6 shows how you can use the LATTICE layout to create Kaplan-Meier graphs for the male and female subgroups. The full code for Program 5-6 can be viewed at https://github.com/rwatson724/SAS-Graphs-Clinical-Trials-Example.

Program 5-6: Creating Subgroup Survival Plot with GTL

```
proc sort data = adam.adtteeff out = adtteeff;
  by sex;   ❶
run;

ods output SurvivalPlot = SurvivalPlot;
ods output HomTests = HomTests(where = (test = "Log-Rank"));
proc lifetest data = adtteeff plots = survival(atrisk = 0 to 210 by 30);
  by sex;   ❶
  time aval * cnsr(1);
  strata trtpn;
run;

/* Obtaining hazard ratios from Proc PHREG */
ods output HazardRatios = HazardRatios;
proc phreg data = adtteeff;
  by sex;   ❶
  class trtpn /ref = last;
  model aval * cnsr(1) = trtpn;
  hazardratio trtpn /diff = ref;
  format trtpn trt.;
run;
```

```
/* Obtaining Interaction p-value from Proc PHREG */
ods output ModelANOVA = Hazard_IntP (where = (EFFECT = "TRTPN*SEX"));   ❷
proc phreg data = adtteeff;
    class trtpn sex /ref = last;
    model aval * cnsr(1) = trtpn|sex;
    format trtpn trt.;
run;

/* create Macro Variables to display hazard ratios and log-rank p-value */
proc sql;
    select put(ProbChiSq, PVALUE6.4) into
                   :log_rank_pvalue1-:log_rank_pvalue2   ❸
    from HomTests;
quit;

proc sql;
    select put(HazardRatio, 6.2) into: HazardRatio1-:HazardRatio4   ❸
    from HazardRatios;
quit;

proc sql;
    select strip(put(ProbChiSq, 6.2)) into: IntP   ❹
    from Hazard_intp;
quit;

proc template;
    define statgraph kmtemplate;
        mvar log_rank_pvalue1 log_rank_pvalue2 HazardRatio1
              HazardRatio2 HazardRatio3  HazardRatio4 IntP;   ❺
        begingraph;

            layout lattice  /columns = 1 rows = 2 columndatarange=union
                rowgutter = 10px;   ❻

            sidebar / align = top;   ❼
                entry textattrs = (weight = bold) "Interaction "
                        textattrs = (style = italic weight=bold) "p"
                        textattrs=(style=normal weight=bold) "-value = " IntP /
                           border = true;   ❽
            endsidebar;

            sidebar / align = left;   ❼
                      entry "Survival Probability" / rotate=90;   ❽
            endsidebar;

            sidebar / align = bottom;   ❼
                discretelegend "Survival1" / across = 3 down = 1;
            endsidebar;

            columnaxes;
                    columnaxis / label = "Days from Randomisation"
                  linearopts = (tickvaluesequence = (start = 0 end = 210
                                   increment = 30));
            endcolumnaxes;

            cell;
                cellheader; ❾
```

```
         entry "Sex='Female'";
      endcellheader;

      layout overlay / yaxisopts=(display=(line ticks
                                           tickvalues));

         stepplot x = time y = eval(ifn(sex = "F",survival,.)) ❿ /
            group = stratum name="Survival1"
            legendlabel="Survival";

         scatterplot x = time y = eval(ifn(sex = "F",censored,.)) /
            markerattrs=(symbol = plus color = black)
            name = "Censored1"
            legendlabel="Censored";

         scatterplot x = time y = eval(ifn(sex = "F",censored,.))/
            markerattrs = (symbol = plus) group = stratum;

         discretelegend "Censored1" / location = inside
            autoalign = (topright);

      << SAS CODE for log-rank and hazard ratio table for females >>

      endlayout;

   endcell;


   cell;
      cellheader;    ❾
         entry "Sex='Male'";
      endcellheader;

      layout overlay / yaxisopts=(display=(line ticks tickvalues));

         stepplot x = time y = eval(ifn(sex = "M",survival,.))   ❿ /
            group = stratum name = "Survival2"
            legendlabel = "Survival";

         scatterplot x = time y = eval(ifn(sex = "M",censored,.))/
            markerattrs = (symbol = plus color = black)
            name = "Censored2"
            legendlabel = "Censored";
         scatterplot x = time y = eval(ifn(sex = "M",censored,.))/
            markerattrs=(symbol = plus) group = stratum;

         discretelegend "Censored2" / location = inside
            autoalign = (topright);

      << SAS CODE for log-rank and hazard ratio table for males >>

      endlayout;

   endcell;

endlayout;
```

```
        endgraph;
    end;
run;

proc sgrender data = Survivalplot template = kmtemplate;
    format stratum $trt.;
run;
```

❶ The PROC SORT, LIFETEST, and PHREG are using the BY statement to produce the results for each gender.

❷ The PROC PHREG is outputting the hazard ratio interaction *p*-value. The reviewers use this *p*-value to assess whether the hazard ratio effects are consistent for each gender. A *p*-value of less than 5% generally indicates that the treatment effects seen are not consistent for both genders.

❸ The macro variables *LOG_RANK_PVALUE1* and *LOG_RANK_PVALUE2* store the log-rank *p*-values for each gender. Since the HOMTESTS data set is sorted alphabetically by gender, the log-rank *p*-value of the females is stored in *LOG_RANK_PVALUE1*, and the males is stored in *LOG_RANK_PVALUE2*. The macro variables *HAZARDRATIO1- HAZARDRATIO4* store the hazard ratios. *HAZARDRATIO1* and *HAZARDRATIO2* contain the hazard ratios of the high dose against the placebo, and the low dose against the placebo, respectively for the female patients. *HAZARDRATIO3* and *HAZARDRATIO4* contain the treatment comparisons for the male patients.

❹ The macro variable *INTP* stores the hazard ratio interaction *p*-value.

❺ The MVAR statement stores the values of the seven macro variables in a character format, ready to be used in the KMTEMPLATE template when called.

❻ The LATTICE layout produces a graph container that has two rows and one column. This enables the Kaplan-Meier curves of the females and males to appear on the same plot. The ROWGUTTER=10PX creates a space of 10px between the two cells. 10px means 10 pixels. Other size dimensions can also be used, such as centimeters, inches, millimeters, percentage, and pt size. These dimensions can be used by specifying CM, IN, MM, PCT or %, and PT respectively, following the quantity.

❼ The SIDEBAR option is useful for displaying information that applies to all of the columns or all of the rows within the graph. The SIDEBAR with the ALIGN=TOP option helps to display the treatment with gender (TRTPN|SEX) interaction *p*-value. The SIDEBAR with the ALIGN=LEFT option helps to create and display the label for the Y axis. This method is used so that there is only one Y-axis label value, instead of one Y-axis label for each row. However, to do this, you need to make sure that the label is not displayed on the Y axis for each individual cell. The SIDEBAR with the ALIGN=BOTTOM option helps to show the treatment legend.

❽ The ENTRY statement is used to display text within the graph. For example, in the first SIDEBAR statement, it is used to display the interaction *p*-value. It can also display the value of a macro variable that is specified with MVAR. Like other GTL statements, you can control the appearance of the text using the different options associated with the TEXTATTRS option. Another feature that can be implemented is to add a box around the text with the BORDER = TRUE option. If you want the text to display vertically, you can use the ROTATE option to indicate the degree of rotation of the text.

⑨ The CELLHEADER statement helps to display the subgroup value at the top of the cell, that is, a description for each subgroup. If you recall, with GTL you can produce single-cell and multi-cell layouts. When you use the LAYOUT statement, you can think of the CELL statement being wrapped around the LAYOUT statement by default. In other words, the LAYOUT statement is nested within the CELL statement. Although when you use the CELLHEADER statement, you need to be explicit and include the CELL and ENDCELL statements around the LAYOUT statement.

⑩ The EVAL and IFN functions are only used to show the Kaplan-Meier results of the female patients in the first cell. This is because when the SEX variable equals "F", the Y value in the STEPPLOT is assigned to the SURVIVAL value, otherwise the Y value is assigned to a missing value. IFN is a short-hand approach to the IF-THEN-ELSE statements. It evaluates the logic in the first argument and either returns the value in the second argument if true or returns the value in the third argument if false. Note that IFN indicates the value that is returned so that it will return a numeric value. For example, for the Y axis in the first cell, IFN was used to return the numeric SURVIVAL values when the SEX was equal to "F", otherwise (numeric) missing values were returned. In the second cell, the EVAL and IFN functions were only used to show the Kaplan-Meier results of the male patients.

Output 5-6 shows the median survival time value for each treatment, separately for female and male patients. You can see from Kaplan-Meier curves, and the interaction *p*-value of 0.31, that the treatment effects are similar between the genders. That is, both male and female patients survive longer when on the high dose of treatment.

Output 5-6: GTL Subgroup Survival Plot by Gender

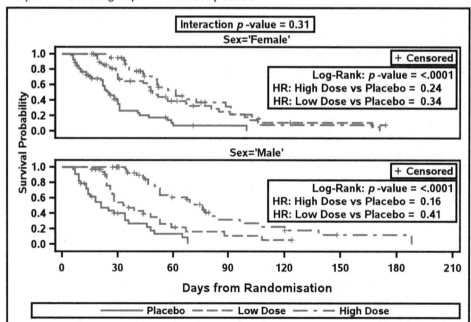

5.3 Forest Plots from Overall Survival Hazard Ratios

Generally, customers request two types of forest plots: forest plots to compare meta-analyses and forest plots to compare subgroup analysis. Meta-analyses involve combining data from multiple studies and analyzing the treatment effect. The forest plots help to display the treatment effect by showing a ratio. Depending on the data, the ratio could be a hazard ratio, odds ratio, or risk ratio. In this section, you will learn how to do forest plots based on the hazard ratio, although you can still apply the same principles to the odds ratio and risk ratio once the ratios are calculated. In the Kaplan-Meier examples in Section 5.2 *Survival Curves for Time-to-Event Endpoints*, you saw how PROC PHREG could be used to output the hazard ratios. In Program 5-6, you also saw how PROC PHREG could be used to output the hazard ratio interaction *p*-value, which is useful when showing subgroups on the forest plot. Once you calculate the hazard ratios, you can essentially use the HIGHLOWPLOT within GTL to produce the forest plots. If you want to produce subgroup forest plots, then it is best to use the LATTICE LAYOUT because you have more control over the layout of the graph. For example, you can create a section that shows the subgroups and subgroup headers, and furthermore, you can add indention to the subgroup values. You can create several column sections, and in one column, you could display the hazard ratios, in another column, you can display the number of patients in the subgroup, and in another column, you could display the number of events in the subgroup.

Display 5-3 shows the hazard ratios and 95% confidence intervals for each subgroup, including the overall subgroup. The first four variables (excluding the OBS variable) are the minimum variables required for you to produce a forest plot. The VARIABLE column contains the factors (subgroup headers) and the subgroup values. The WALDLOWER and WALDUPPER contain the lower and upper 95% confidence intervals. The INDENT column is used to add an indention to the VARIABLE value in the forest plot. For example, a value of 0 indicates that no indention is to be added. The SUBGROUP_HEADER column defines the subgroup headings, and this is used to format the subgroup headings with the bold format.

Display 5-3: Data Table of Hazard Ratios

Obs	Variable	HazardRatio	WaldLower	WaldUpper	Indent	Subgroup_Header
1	Overall	0.203	0.138	0.298	0	Yes
2	Age	.	.	.	0	Yes
3	<65	0.227	0.085	0.603	1	No
4	65-80	0.230	0.139	0.382	1	No
5	>80	0.135	0.057	0.320	1	No
6	Race	.	.	.	0	Yes
7	White	0.196	0.130	0.295	1	No
8	Black or African American	0.231	0.067	0.801	1	No
9	Sex	.	.	.	0	Yes
10	Female	0.240	0.142	0.407	1	No
11	Male	0.156	0.087	0.281	1	No

When you do subgroup analysis, it is essential to have variables that will order your data so that the final data set which the plot uses is in the correct sequence. You will need a variable to order your subgroup headers, and another one to order your subgroup values. Program 5-7 shows how you can produce a data set that contains the hazard ratios and necessary subgroups in the correct order, and Program 5-8 shows how you can produce a subgroup forest plot. You can reference the complete code for Program 5-7 at https://github.com/rwatson724/SAS-Graphs-Clinical-Trials-Example.

Program 5-7: Creating Data Set for Subgroup Forest Plot

```
<< SAS CODE for creating desired subgroup labels using PROC FORMAT>>   ❶

<< SAS CODE for ordering subgroup as desired using PROC FORMAT>>   ❷

data adtteeff;
   set adam.adtteeff;
   overall = 1;   ❸
run;

options mprint;
%macro subgroup(variable=, displayvariable=, fmt=, fmt2=, ord=);   ❹

   proc sort data = adtteeff;
      by &variable.;   ❺
   run;

   ods output HazardRatios = HazardRatios_&variable.;
   proc phreg data = adtteeff;
      by &variable.;   ❻
      class trtpn /ref=last;
      model aval * cnsr(1) = trtpn;
      hazardratio trtpn /diff=ref;
```

```
      format trtpn trt.;
   run;

   data HazardRatios_&variable.2(drop = &variable.);
      set HazardRatios_&variable.;
      length variable $30;
      variable = put(&variable., &fmt..);    ❻
      original_order = put(variable, &fmt2..);    ❼
      ord = &ord.;
      if variable = "Overall" then ord2 = 1;
      else ord2 = 2;
   run;

   << SAS CODE to create variable headers >>

   data headers;    ❽
      set %if &ord > 2 %then headers;
         headers_&variable.;
   run;

   data HazardRatios_Subgroup;    ❾
      set %if &ord > 1 %then HazardRatios_Subgroup;
         HazardRatios_&variable.2;
   run;

%mend;
%subgroup(variable=overall, fmt=overfmt, fmt2=$overfmto, ord=1);
%subgroup(variable=agegr1n, displayvariable=Age, fmt=agefmt,
         fmt2=$agefmto, ord=2);
%subgroup(variable=racen, displayvariable=Race, fmt=racefmt,
         fmt2=$racefmto, ord=3);
%subgroup(variable=sex, displayvariable=Sex, fmt=$sexfmt,
         fmt2=$sexfmto, ord=4);

data HazardRatios_Subgroup_All;    ❿
   set HazardRatios_Subgroup headers;
   if ord2 = 1 then do;
      indent = 0;
       subgroup_header = "Yes";
   end;
   else do;
      indent = 1;
       subgroup_header = "No";
   end;
run;
```

❶ PROC FORMAT helps to display the desired subgroup values on the plot.

❷ PROC FORMAT helps to order the subgroup values in the desired order.

❸ An overall variable is created with just one value, so that a macro can also be used to analyze the overall group. For the purpose of this example, "Overall" represents all the patients.

❹ The macro SUBGROUP is used to calculate the hazard ratio for each subgroup and put all the hazard ratios into one data set. The SUBGROUP macro has five macro parameters. These are *VARIABLE, DISPLAYVARIABLE, FMT, FMT2,* and *ORD*.

- The *VARIABLE* macro variable requires the factor that you want to analyze, and which is in the analysis data set, for example, AGEGR1N.

- In the *DISPLAYVARIABLE* macro variable, you will indicate the text that you want displayed on the graph. For example, you might want to display "Age group" instead of "AGEGR1N".

- In the *FMT* macro variable, you will specify your desired format. The *FMT* macro variable is used to show the subgroup values in the format that you require. For example, you might want to display "Female" instead of the value "F".

- The *FMT2* macro variable helps to order the subgroups per the specified requirements. For example, you might want to have "Males" before "Females" or vice versa.

- The *ORD* macro variable helps to order the subgroups per the specified requirements. For example, you might want to display the subgroups relating to RACE before AGEGR1N.

❺ The PROC SORT and PHREG are using the BY statement to produce the results for each factor in the macro call.

❻ The PUT function outputs the subgroup value in the desired format and stores it in a new variable named VARIABLE.

❼ The PUT function creates an order variable named ORIGINAL_ORDER so that the subgroups can be ordered as requested.

❽ The different factors of the analysis are all put into the HEADERS data set in the correct order.

❾ The subgroups are all put into the HAZARDRATIOS_SUBGROUP.

❿ The data set containing the factors and hazard ratios of the subgroups is created. This data set is called HAZARDRATIOS_SUBGROUP_ALL and it is the data set that is used to generate the forest plot.

Display 5-4 is a snippet of the data set HAZARDRATIOS_SUBGROUP_ALL sorted by the variables ORD, ORD2 and ORIGINAL_ORDER and filtered by the hazard ratios and statistics belonging to the high dose versus the placebo group. The DESCRIPTION column indicates which hazard ratios and statistics belong to the corresponding treatment versus placebo group has been removed from this data snippet to save space. For the purpose of the data snippet, DESCRIPTION is "TRTPN High Dose vs Placebo". The sort variables order the data set appropriately by first displaying the Overall subgroup header (VARIABLE), which uses all the data, and then the Age, Race and Sex subgroup headers (SUBGROUP_HEADER). Also, within the subgroup headers, the subgroup values are sorted as required. For example, within Age, the order of the subgroup values is <65, followed by 65-80 and then followed by >80.

Display 5-4: Data Table of Hazard Ratios with Sorting Variables (HAZARDRATIOS_SUBGROUP_ALL)

Obs	HazardRatio	WaldLower	WaldUpper	Variable	Original_order	Ord	Ord2	Indent	Subgroup_Header
1	0.203	0.138	0.298	Overall	1	1	1	0	Yes
2	.	.	.	Age		2	1	0	Yes
3	0.227	0.085	0.603	<65	1	2	2	1	No
4	0.230	0.139	0.382	65-80	2	2	2	1	No
5	0.135	0.057	0.320	>80	3	2	2	1	No
6	.	.	.	Race		3	1	0	Yes
7	0.196	0.130	0.295	White	1	3	2	1	No
8	0.231	0.067	0.801	Black or African American	2	3	2	1	No
9	.	.	.	Sex		4	1	0	Yes
10	0.240	0.142	0.407	Female	1	4	2	1	No
11	0.156	0.087	0.281	Male	2	4	2	1	No

Now that you know how to produce a data set for the subgroup analysis, you will learn how to produce the forest plot. Program 5-8 shows how you can produce a subgroup forest plot.

Program 5-8: Creating Subgroup Forest Plot

```
proc sort data = tfldata.HazardRatios_Subgroup_All out =
HazardRatios_Subgroup_All;
   by ord ord2 original_order;   ❶
run;

proc template;
   define statgraph forestplotsubgrouptemplate;
      begingraph;

         discreteattrmap name = "factortext";   ❷
            value "Yes" / textattrs = (weight = bold);
            value "No" / textattrs =(weight = normal);
         enddiscreteattrmap;
         discreteattrvar attrvar = factortext var = factor
                     attrmap = "factortext";

         layout overlay / xaxisopts = (type = log
                           logopts = (base = 2
                                      tickintervalstyle = logexpand
                                      viewmin= 0.0625 viewmax = 16)   ❸
                           label="Hazard Ratio"
                           tickvalueattrs=(size = 8pt)
                           labelattrs = (size = 9pt))
                        yaxisopts = (reverse = true display = (line)
                           label ="Interventions"
                           tickvalueattrs=(size=8pt)
                           labelattrs=(size=9pt));   ❸

            innermargin / align = left;   ❹
               axistable y = variable value = variable / display = (values)
```

```
                    indentweight = indent textgroup = factortext;  ❺
            endinnermargin;

            highlowplot y = variable low = WaldLower high = WaldUpper;  ❻

            scatterplot y = variable x = HazardRatio /
                        markerattrs = (symbol = squarefilled);  ❼

            referenceline x = 1 / lineattrs = (pattern = 2);  ❽

            layout gridded/border = false halign = left valign = bottom;  ❾

                entry haling = left "Xanomeline high dose is better" /
                    textattrs = (size = 5);

            endlayout;

            layout gridded/border = false haling = right valign = bottom;  ❾

                entry halign=left "Placebo is better" /
                    textattrs = (size = 5);

            endlayout;
        endlayout;
      endgraph;
    end;
run;

proc sgrender data = Hazardratios_subgroup_all
            template = forestplotsubgrouptemplate;
    where description = "TRTPN High Dose vs Placebo";❿
run;
```

❶   PROC SORT sorts the data appropriately. When creating a forest plot the order of the data is important. You might not always be able to rely on the value of the data used to produce the plot. Thus, creating one or more order variables to help maintain the desired order might be necessary.

❷   As discussed in Section 2.5.5 *Producing Frequency Plot of Adverse Events Grouped by Maximum Grade for Each System Organ Class*, an attribute map enables you to assign specific attributes to various graphic components based on the data value. In this particular example, the attribute map is created so that the subgroup headers are displayed in bold text while the subgroup values are displayed using the normal font-weight. Based on the definition of attribute map if a record within the data set has FACTOR = "Yes" then the variable value will be bold and if the record has FACTOR = "No" then the variable value will have normal font-weight.

❸   Like all the other examples up to this point, the axes had special options to help control the appearance. For the forest plot, additional options are introduced. For this example, in order to make the hazard ratios appear symmetrical, the following options are used TYPE = LOG and LOGOPTS in the X-axis options. Within LOGOPTS there are additional options that can be specified; BASE = 2 and TICKINTERVALSTYLE = LOGEXPAND. Otherwise, one of the 95% confidence intervals will appear wider than the other because the default TYPE = LINEAR axis option will be used. In some cases, it might be good to find the absolute maximum value and then find a value in terms of base 2, which is bigger than the absolute maximum. Then use that value for both the minimum and maximum values on the X axis so that the X axis is symmetrical around the value of 1. For example, the minimum value is 2^{-4} and the maximum value is 2^4, that is, the minimum value is 0.0625 and the maximum value is 16. If you have examples, where the hazard ratios are relatively large, such as in the hundreds or thousands, or relatively small, then using BASE = 10 could be more appropriate. Note that using the LOGOPTS option is not required if you are not using TYPE = LOG. In other words, if you are creating a forest plot that has TYPE = LINEAR, then you do not need to worry about rescaling the X axis.

An additional option that you need to use is the REVERSE = TRUE, which is illustrated on the YAXISOPTS. On the Y axis the data is displayed in increasing order from bottom to top. However, for a forest plot, it needs to be displayed in increasing order from top to bottom. Thus, when creating forest plots, it is important to use the REVERSE = TRUE on the Y-axis option. Otherwise, the subgroup values will be displayed in the opposite order then what is intended.

❹ In this example, the INNERMARGIN statement creates space on the left side of the plot for the factors and subgroup values. It is also possible to create space on the right-hand side of the plot by using the ALIGN = RIGHT option in the INNERMARGIN statement.

❺ The AXISTABLE statement displays the factor and subgroup values. Because you want the values to be displayed on the Y axis, you can assign the Y argument to the variable that contains the value.

- The VALUE option indicates the values to use on the plot.

- The INDENTWEIGHT option indicates how much indention each of the values should have. The subgroup headers have an indention of 0, and the subgroup values have an indention of 1. One indent is equivalent to $1/8^{th}$ of an inch indention.

- The TEXTGROUP = FACTORTEXT is used to obtain the discrete attribute map instructions and hence, display bold text for the factors.

❻ The HIGHLOWPLOT statement draws a line from the lower 95% hazard ratio confidence intervals to the upper 95%. This essentially describes the hazard ratio for each subgroup.

❼ The SCATTERPLOT statement indicates the actual hazard ratio on the plot with a square filled marker.

❽ The REFERENCELINE statement draws a dashed vertical line where the hazard ratio equals one. This dashed vertical line can be used as an indicator of a treatment difference or not. The dashed vertical line can be used as an indicator of treatment difference because if the confidence intervals cross this line, then this indicates that there is no evidence to suggest a treatment difference. However, if both the lower and upper confidence limits are on one side of the reference line then this indicates that there is a treatment difference.

❾ The GRIDDED layout helps to position the text, which indicates the conclusions of the hazard ratio. Hazard ratios toward the left-hand side of the graph indicate that the high dose is better, and hazard ratios toward the right-hand side indicate that the placebo is better. The arrows at the bottom left and bottom right side of the graph can assist with the conclusions of the hazard ratios.

⑩ The data set has been subsetted to only show the hazard ratios of the high dose versus placebo.

Output 5-7 shows the forest plot of the subgroups for overall survival. The forest plot of the subgroups indicates that the treatment effects are quite consistent regardless of the subgroup used.

Output 5-7: Subgroup Forest Plots of Overall Survival

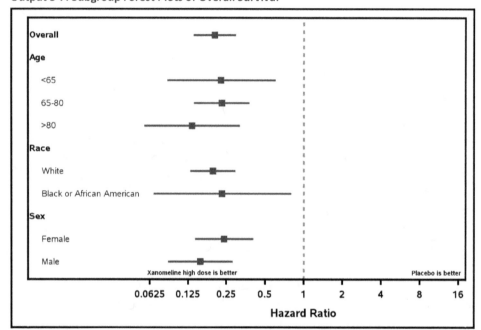

5.3.1 Generating Forest Plot that Exhibit Number of Observations

In Output 5-7, all of the square symbols that represent the hazard ratio are the same size. It is common for the size of the symbols to represent the number of observations that went into creating the hazard ratio. One method that you can use to make the size of the symbols represent the number of observations is to use the SIZERESPONSE option. Another method that you can use is the SIZE option. If you only need to make one stand-alone plot containing the observation size within a symbol, you can use the SIZERESPONSE option.

However, if you would like to be able to compare the symbol size between plots, you should either use the SIZE option, or ensure that all plots have the same minimum and maximum value that represents the symbol size before using the SIZERESPONSE option. For example, if you have two different responses, such as Treatment A and Treatment B, and you want to display a symbol that represents the number of patients, so that when the number of patients is high the symbol will be large and when the number of patients is low the symbol will be small and you want to use the SIZERESPONSE option to display the magnitude and you want to also be able to compare the sizes of the symbol, then you should ensure that both treatments have the same minimum and maximum value. If Treatment A has the minimum number of patients as 100 and the maximum number of patients as 5000, and Treatment B has the minimum and maximum number of patients as 10 and 20 respectively, then SIZERESPONSE could not be used in this situation because the minimum and maximum values for the two treatments are not the same. If you were to use the SIZERESPONSE option, and output both treatment plots, then the size of the symbol that represented 100 patients in Treatment A would also be the same size of the patients that represented 10 patients in Treatment B. Therefore, if you wanted to compare between plots, then in Treatment A, you would need to add another record in the data set that had a value that indicated that the number of patients is 10, and in Treatment B you would need to add an additional record in the data set that indicated that there are 5000 patients. Both records that you add would contain blank values in the X or Y variables, so these values would not actually be plotted.

Adding these dummy records enables you to compare the sizes between plots because the symbol sizes would be the correct size. The reason you need to make sure that the minimum and maximum values are the same is because the minimum and maximum establish a range over which the marker sizes vary in linear proportion (SAS Institute Inc., 2020). This section demonstrates how to use the SIZE option when comparing symbol sizes between plots.

Display 5-5 is a snippet of the data set HAZARDRATIOS_SUBGROUP_ALL_N, and Display 5-5 is very similar to Display 5-4, except that it contains an additional column with the number of observations in the hazard ratio. If you recall, there are 86 placebo patients and 84 patients on xanomeline high dose. Therefore, the hazard ratio comparison of high dose vs placebo contains 170 patients and that is reflected with a TOTAL_N value of 170 for the Overall comparison.

Display 5-5: Data Table of Hazard Ratios with Number of Observations (HAZARDRATIOS_SUBGROUP_ALL_N)

Obs	HazardRatio	WaldLower	WaldUpper	Variable	Original_order	Ord	Ord2	Indent	Subgroup_Header	Total_N
1	0.203	0.138	0.298	Overall	1	1	1	0	Yes	170
2	.	.	.	Age		2	1	0	Yes	.
3	0.227	0.085	0.603	<65	1	2	2	1	No	25
4	0.230	0.139	0.382	65-80	2	2	2	1	No	97
5	0.135	0.057	0.320	>80	3	2	2	1	No	48
6	.	.	.	Race		3	1	0	Yes	.
7	0.196	0.130	0.295	White	1	3	2	1	No	152
8	0.231	0.067	0.801	Black or African American	2	3	2	1	No	17
9	.	.	.	Sex		4	1	0	Yes	.
10	0.240	0.142	0.407	Female	1	4	2	1	No	93
11	0.156	0.087	0.281	Male	2	4	2	1	No	77

Program 5-9 shows how you can create a forest plot with the marker symbol size represented by the number of patients.

Program 5-9: Creating Subgroup Forest Plot with Number of Observations as Marker Symbol Size

```
proc sort data = HazardRatios_Subgroup_All_N;
   by ord ord2 original_order;   ❶
run;

proc template;
   define statgraph forestplotsubgrouptemplate;
      begingraph;

         discreteattrmap name = "factortext";
            value "Yes" / textattrs = (weight = bold);
            value "No" / textattrs = (weight = normal);
         enddiscreteattrmap;

         discreteattrvar attrvar = factortext var = factor
                         attrmap = "factortext";

         layout overlay / xaxisopts = (type = log
                          logopts = (base = 2
                                          tickintervalstyle = logexpand
                              viewmin = 0.0625 viewmax = 16)
                          label = "Hazard Ratio"
                          tickvalueattrs = (size = 8pt)
                          labelattrs = (size = 9pt))
                        yaxisopts = (reverse = true display = (line)
                          label = "Interventions"
                          tickvalueattrs = (size = 8pt)
                          labelattrs = (size = 9pt));

            innermargin / align = left;
               axistable y = variable value = variable / display = (values)
                  indentweight = indent textgroup = factortext;
            endinnermargin;

            highlowplot y = variable low = WaldLower high = WaldUpper;

            scatterplot y = variable x = HazardRatio /
                        markerattrs = (symbol = squarefilled)
                        sizeresponse = total_n;   ❷

            referenceline x = 1 / lineattrs = (pattern = 2);

            layout gridded / border = false halign = left valign = bottom;

               entry haling = left "Xanomeline high dose is better" /
                  textattrs = (size = 5);

            endlayout;

            layout gridded / border = false haling = right valign = bottom;

               entry halign=left "Placebo is better" /
                  textattrs = (size = 5);
```

```
        endlayout;

        endlayout;
      endlayout;
    endgraph;
  end;
run;

proc sgrender data = Hazardratios_subgroup_all_n template =
forestplotsubgrouptemplate;
   where description = "TRTPN High Dose vs Placebo";
run;
```

❶ PROC SORT sorts the HAZARDRATIOS_SUBGROUP_ALL_N data set appropriately. As mentioned in Program 5-8, when creating a forest plot the order of the data is important. You might not always be able to rely on the value of the data used to produce the plot. Thus, creating one or more order variables to help maintain the desired order might be necessary.

❷ You can use the SIZERESPONSE option to change the size of your marker symbol based on the values within the assigned column.

Output 5-8 shows the forest plot of overall survival for some subgroups, and you can see that the size of the hazard ratio symbol varies across the subgroups. The larger the hazard ratio symbol size, the larger the number of patients. Understandably, the marker symbol for the overall subgroup is the largest because it contains the greatest number of patients: 170. When you look at the graph, it is apparent that more patients over the age of 80 contributed to deriving the hazard ratio compared to patients younger than 65.

Output 5-8: Subgroup Forest Plots of Overall Survival Plot with Number of Observations as Marker Symbol Size

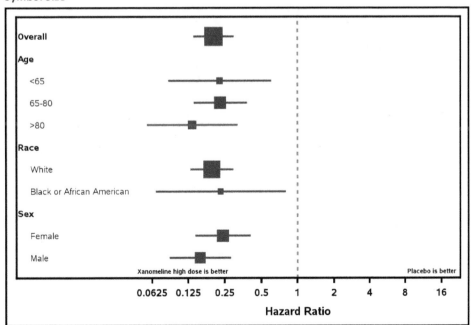

5.3.2 Generating Multiple Forest Plots that Exhibit Number of Observations

In Output 5-8, you saw how the marker size varied depending on the number of observations used for the hazard ratios that compared the high dose to the placebo. If you wanted to also see how the marker size varied for the low dose versus the placebo comparison and you wanted to also be able to compare the marker sizes between the low dose and high dose, then you can use the SIZE option to do this.

Program 5-10 shows you how you can create multiple forest plots with the marker symbol size represented by the number of patients and compare between forest plots. You can view Program 5-10 in its entirety at https://github.com/rwatson724/SAS-Graphs-Clinical-Trials-Example.

Program 5-10: Creating Multiple Subgroup Forest Plots with Number of Observations as Marker Symbol Size

```
proc sort data = tfldata.HazardRatios_Subgroup_All_N
   out = HazardRatios_Subgroup_All_N;
   by ord ord2 original_order;   ❶
run;

<< SAS CODE to produce macro variables for low and high values >>  ❷ ❸

proc template;
   define statgraph forestplotsubgrouptemplate;
   dynamic dose;   ❹
   mvar high_n1 high_n2 high_n3 high_n4 high_n5 high_n6 high_n7 high_n8  ❷
        low_n1 low_n2 low_n3 low_n4 low_n5 low_n6 low_n7 low_n8;   ❸
      begingraph;

         discreteattrmap name = "factortext";
            value "Yes" / textattrs = (weight = bold);
            value "No" / textattrs =(weight = normal);
         enddiscreteattrmap;
         discreteattrvar attrvar = factortext var = factor
                     attrmap = "factortext";

         layout overlay / xaxisopts=(type=log
                           logopts = (base = 2
                                      tickintervalstyle = logexpand
                           viewmin = 0.03125 viewmax = 32)
                           label = "Hazard Ratio"
                           tickvalueattrs = (size = 8pt)
                           labelattrs=(size=9pt))
                        yaxisopts = (reverse = true display = (line)
                           label="Interventions"
                           tickvalueattrs=(size=8pt)
                           labelattrs=(size=9pt));

            innermargin / align=left;
               axistable y = variable value = variable / display = (values)
                  indentweight = indent textgroup = factortext;
            endinnermargin;

         highlowplot y = variable low = WaldLower high = WaldUpper;
```

```
            if (dose = "HIGH")    ❹

                scatterplot y = variable
                    x = eval(ifn(variable = "Overall", HazardRatio, .)) /   ❺
                        markerattrs = (symbol = squarefilled size = high_n1);   ❺

    << Repeat SCATTERPLOT for each unique value of AGEGR1N, RACE, GENDER >>

            else

                scatterplot y = variable ❻
                    x = eval(ifn(variable = "Overall", HazardRatio, .)) /
                        markerattrs = (symbol = squarefilled size = low_n1);   ❻


    << Repeat SCATTERPLOT for each unique value of AGEGR1N, RACE, GENDER >>

            endif;

            referenceline x = 1 / lineattrs = (pattern = 2);

            layout gridded / Border = false haling = left valign = bottom;
                layout gridded;
                  if (dose="HIGH")    ❼
                    entry haling = left "Xanomeline high dose is better" /
                        textattrs=(size=5);
                  else
                    entry halign=left "Xanomeline low dose is better" /
                        textattrs=(size=5);
                    endif;
                endlayout;
            endlayout;

        << SAS CODE for "Placebo is better" text >>

            endlayout;
        endgraph;
    end;
run;


proc sgrender data = Hazardratios_subgroup_all_n ❽
    template = forestplotsubgrouptemplate;
        dynamic dose = "HIGH";   ❾
        where description = "TRTPN High Dose vs Placebo";
run;
proc sgrender data = Hazardratios_subgroup_all_n ❿
    template = forestplotsubgrouptemplate;
        where description = "TRTPN Low Dose vs Placebo";
run;
```

❶ PROC SORT sorts the HAZARDRATIOS_SUBGROUP_ALL_N data set appropriately.

❷ You need to determine the total number of patients in the hazard ratios that compare the high dose to the placebo for each of the subgroups and store the values in macro variables. The total number is divided by a constant amount of 7 to make the size smaller than the original size, and to make the symbol negligible when the number of observations is close to 0.

❸ Similar to what was done for the high dose, you also need to determine total number of patients in the hazard ratios that compare the low dose to the placebo for each subgroup and store the values in macro variables and divide the total number by a constant amount of 7 to make the size smaller and to make the symbol negligible when the number of observations is close to 0.

❹ A dynamic variable called *DOSE* is used in conjunction with conditional logic to determine which statements to render. If DOSE = "HIGH", then only the number of observations, related to the high dose versus placebo comparison, will be rendered in the template. Otherwise, the template will render the number of observations related to the low dose versus placebo comparison.

❺ Similar to Program 5-6, the EVAL and IFN functions are only used to show the results that meet the conditions. When the conditions are met, a scatter plot of the hazard ratio is produced. You can use the SIZE option within the MARKERATTRS option to control the size of the marker. In this example, the options MARKERATTRS = (SYMBOL = SQUAREFILLED SIZE = HIGH_N1) define a filled-in square as the marker, and that the size of the maker should be HIGH_N1. *HIGH_N1* is a macro variable with the macro variable value equal to the number of observations in the overall subgroup between the high dose versus the placebo comparison.

❻ The statements in the section following the ELSE statement are related to the low dose versus the placebo comparison. In other words, when the dynamic variable is not equal to "HIGH", the SCATTERPLOT statements associated with the low dose versus placebo are used.

❼ The conditional IF block determines whether the plot should display "Xanomeline high dose is better" or "Xanomeline low dose is better". When DOSE = "HIGH", the plot displays "Xanomeline high dose is better".

❽ The SGRENDER code creates the plot of hazard ratios for the high dose versus the placebo comparison.

❾ DYNAMIC DOSE = "HIGH" creates a dynamic variable called DOSE, with a value of "HIGH". For the dynamic variable to work, it also needs to be assigned within the template, as was done in callout ❹.

❿ The SGRENDER code creates the plot of hazard ratios for the low dose versus the placebo comparison.

Output 5-9 shows the forest plots of the overall survival hazard ratios for both the high dose versus the placebo comparisons and the low dose versus the placebo comparisons. In Output 5-9, you can directly compare the sizes of the hazard ratios between the two plots.

Output 5-9: Subgroup Forest Plots of Overall Survival Plot with Number of Observations as Marker Symbol Size

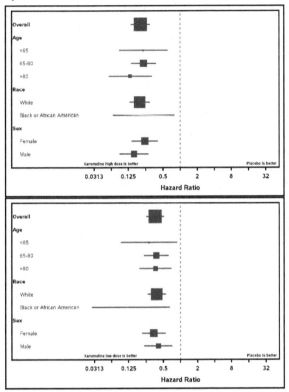

5.4 Chapter Summary

Depending on the information that is to be captured on the Kaplan-Meier Curves and forest plots, you might need to first calculate the necessary statistics. With the use of various GTL plot statements such as the STEPPLOT, SCATTERPLOT, and AXISTABLE statements and text statements such as the ENTRY statement the desired output can be produced to help with the generation of the visual representation of the results.

Kaplan-Meier curves on overall survival and progression-free survival are useful for assessing treatment efficacy, and when combined with log-rank *p*-values, median survival times, and hazard ratios, they can be even more valuable because all the necessary information is provided on one figure. Forest plots of the subgroups are useful for investigating if the treatment effects are consistent for each subgroup, and to investigate if the treatment effects are significant.

References

Matange, S. (2014, February 09). "Survival Plot." Retrieved from SAS Blogs: Graphically Speaking:
https://blogs.sas.com/content/graphicallyspeaking/2014/02/09/survival-plot/

Meyers, J. (2018, 07 20). "Kaplan-Meier Survival Plotting Macro %NEWSURV." Retrieved from SAS Communities:
https://communities.sas.com/t5/SAS-Communities-Library/Kaplan-Meier-Survival-Plotting-Macro-
NEWSURV/ta-p/479747

Pocock, S. J., Clayton, T. C., and Altman, D. G. (2002). "Survival plots of time-to-event outcomes in clinical trials: good
practice and pitfalls." *The Lancet*, v.359, no.9318, p.1686-1689.

SAS Institute Inc. (2020, 07 29). *SAS® 9.4 Graph Template Language: Reference, Fifth Edition*. Retrieved from SAS® 9.4
and SAS® Viya® 3.5 Programming Documentation: https://documentation.sas.com/:
https://documentation.sas.com/?docsetId=grstatgraph&docsetTarget=p08qhdljvzthmnn1vkbule00ad6r.htm&
docsetVersion=9.4&locale=en#n0jx695r8jau0ln1fym7j8t5rulj

Tanniou, J., van der Tweel, I., Teerenstra, S., and Roes, K. C. B.(2016). "Subgroup analyses in confirmatory clinical trials:
time to be specific about their purposes." *BMC Medical Research Methodology, 16, 20*, 1. Retrieved from
https://doi.org/10.1186/s12874-016-0122-6

U.S. Department of Health and Human Services. (2020, 04 12). (2018) *U.S. Food and Drug Administration*. Clinical Trial
Endpoints for the Approval of Cancer Drugs and Biologics: Guidance for Industry. Retrieved from
https://www.fda.gov/media/71195/download

Wikipedia. (2020, 03 29). "Kaplan–Meier estimator." Retrieved from
https://en.wikipedia.org/wiki/Kaplan%E2%80%93Meier_estimator

Chapter 6: Using SAS to Generate Graphs for Tumor Response

6.1 Introduction on Tumor Response

There are more than 100 different types of cancers, and they are named for the part of the body that is affected. These cancers can be broken into several categories: carcinoma, sarcoma, leukemia, lymphoma, multiple myeloma, melanoma, brain and spinal cord tumors, as well as other types of tumors. (About Cancer: Understanding Cancer, 2015)

One of the most important features in an oncology study for solid tumors is the change in tumor burden. Tumor burden, also known as tumor load, is the amount of cancerous cells that are in the body. Essentially, tumor burden is the size of the tumor. Key endpoints for oncology studies are tumor shrinkage (objective response) and time to disease progression. The objective response is determined by using the Response Evaluation Criteria in Solid Tumors (RECIST) (Eisenhauer et al., 2009). For the purpose of this chapter, tumor burden, tumor load, and tumor size all represent the same concept.

Although there are other types of cancers, this chapter focuses on solid tumors in order to demonstrate how you can use SAS to produce visualizations on changes in tumor burden over time as well as the patient's objective response.

The ADaM data set used for the illustration of graphs using tumor size and response data are the creation of the authors and do not reflect real patient data. The objective response does not align with the RECIST criteria. Rather, it uses the disease response values only to demonstrate their use for producing the visualizations.

6.2 Assessment of Tumor Burden

At baseline, all tumors are either categorized as measurable or non-measurable. Measurable lesions are also known as target lesions, while non-measurable are known as non-target. A tumor is measured using either computed tomography (CT), magnetic resonance imaging (MRI) scan, caliper measurement or chest X-ray. For a tumor to be considered measurable it needs to be measured in one dimension with the minimum size being:

- at least 10 mm if measured by a CT scan

- at least 10 mm if measured by caliper measurement by clinical exam, or

- at least 20 mm if measured by chest X-ray

Regardless of which method is used the longest diameter is recorded and this is used for assessing the tumor burden (Eisenhauer, et al., 2009).

Each tumor is measured at each time point during the course of the study in order to see how the patient is responding to treatment. The standard way to measure the patient's response to treatment is through the use of RECIST.

There are four types of response a patient can have (Eisenhauer et al., 2009):

- Complete Response (CR): disappearance of all target lesions

- Partial Response (PR): at least a 30% decrease from baseline in the sum of diameters of target lesions

- Stable Disease (SD): neither significant shrinkage or increase

- Progressive Disease (PD): at least a 20% increase from baseline in the sum of diameters of target lesions

In some cases, a tumor might not be evaluable (NE), and the tumor measurement might not be used in the assessment. For the purpose of our examples, we assume that all lesions are target lesions and that NE lesions will not be displayed on the outputs.

Using CDISC SDTM standards, tumor identification (TU), tumor results (TR), and disease response (RS) are recorded separately. However, when used for analysis, the data can be combined to produce one or more ADaM compliant data sets to aid in the production of outputs.

Display 6-1 is a snippet of the tumor diameter measurements for each lesion at each visit. One of the criteria for RECIST is to take the sum of the longest diameters across all lesions. Thus, the longest diameter (LDIAM) is data that is collected; however, sum of the longest diameter (SUMLDIAM) is calculated. For example, rows 4 - 6 correspond to the baseline value for the tumor size for each location. The sum of these records is used to produce the analysis value for row 26, that is 4 + 35 + 37 = 76. This value is used as the baseline value and is flagged with ABLFL = "Y" to indicate it is to be used for baseline assessments. Data from the SDTM domain TU can be carried over to the ADaM data set to help with the identification of tumors. In this example, we use the tumor location (TULOC) for tumor identification. In addition, to calculating the parameter SUMLDIAM, new variables are calculated, which will be used to produce the outputs.

The change from baseline (CHG) and percentage change from baseline (PCHG) are used to assess the tumor response. Looking at the SUMLDIAM records that occur post-baseline, you can see that for rows 27–30, there was an increase in the tumor size when compared to the baseline value. This increase is evident with the positive PCHG. However, with rows 31–32, you see a decrease in tumor size based on the negative PCHG. The analysis level flag (ANL01FL) is used to select the record that indicated the best response (that is the smallest percentage change). In Display 6-1, row 31 represents the smallest percentage change and is therefore flagged for analysis. BOR represents the best overall response.

Display 6-1: Data Snippet of Analysis Data Set for Tumor Diameter (ADTUMD)

Obs	USUBJID	PARAMCD	PARAM	TULOC	AVISIT	ADT	ADY
1	01-701-1015	LDIAM	Longest Diameter (mm)	Breast	Screening	3-Oct-13	-91
2	01-701-1015	LDIAM	Longest Diameter (mm)	Lung	Screening	3-Oct-13	-91
3	01-701-1015	LDIAM	Longest Diameter (mm)	Other	Screening	3-Oct-13	-91
4	01-701-1015	LDIAM	Longest Diameter (mm)	Breast	Baseline	26-Dec-13	-7
5	01-701-1015	LDIAM	Longest Diameter (mm)	Lung	Baseline	26-Dec-13	-7
6	01-701-1015	LDIAM	Longest Diameter (mm)	Other	Baseline	26-Dec-13	-7
7	01-701-1015	LDIAM	Longest Diameter (mm)	Breast	Month 1	25-Jan-14	24
8	01-701-1015	LDIAM	Longest Diameter (mm)	Lung	Month 1	25-Jan-14	24
9	01-701-1015	LDIAM	Longest Diameter (mm)	Other	Month 1	25-Jan-14	24
...	...						
25	01-701-1015	SUMLDIAM	Sum of Longest Diameter (mm)		Screening	3-Oct-13	-91
26	01-701-1015	SUMLDIAM	Sum of Longest Diameter (mm)		Baseline	26-Dec-13	-7
27	01-701-1015	SUMLDIAM	Sum of Longest Diameter (mm)		Month 1	25-Jan-14	24
28	01-701-1015	SUMLDIAM	Sum of Longest Diameter (mm)		Month 2	22-Feb-14	52
29	01-701-1015	SUMLDIAM	Sum of Longest Diameter (mm)		Month 3	24-Mar-14	82
30	01-701-1015	SUMLDIAM	Sum of Longest Diameter (mm)		Month 4	22-Apr-14	111
31	01-701-1015	SUMLDIAM	Sum of Longest Diameter (mm)		Month 5	23-May-14	142
32	01-701-1015	SUMLDIAM	Sum of Longest Diameter (mm)		Month 6	18-Jun-14	168

Obs	AVAL	BASE	ABLFL	CHG	PCHG	ANL01FL	BOR
1	19	4		15	375		PR
2	49	35		14	40		PR
3	26	37		-11	-29.7297		PR
4	4	4	Y	0	0		PR
5	35	35	Y	0	0		PR
6	37	37	Y	0	0		PR
7	8	4		4	100		PR
8	39	35		4	11.42857		PR
9	35	37		-2	-5.40541		PR
...							
25	94	76		18	23.68421		PR
26	76	76	Y	0	0		PR
27	82	76		6	7.894737		PR
28	85	76		9	11.84211		PR
29	85	76		9	11.84211		PR
30	80	76		4	5.263158		PR
31	71	76		-5	-6.57895	Y	PR
32	72	76		-4	-5.26316		PR

Display 6-2 is a snippet of the tumor response at each visit. The standard methodology for assessing the change in tumor size (that is the tumor response) is RECIST. The tumor response can be CR, PR, SD, or PD and is captured in the analysis value (AVALC). The analysis level flag (ANL01FL) is used to select the record that indicated the best overall response. Note that the best overall response from ADRESP is also captured in ADTUMD. The purpose of capturing this information in both data sets is demonstrated in the waterfall plot example that uses Display 6-4 and Program 6-3.

Display 6-2: Data Snippet of Analysis Data Set for Tumor Response (ADRESP)

Obs	USUBJID	PARAMCD	PARAM	AVISITN	AVISIT	ADT	ADY	AVALC	ANL01FL
1	01-701-1015	OALLRESP	Overall Response	0	Baseline	26-Dec-13	-7		
2	01-701-1015	OALLRESP	Overall Response	1	Month 1	25-Jan-14	24	SD	
3	01-701-1015	OALLRESP	Overall Response	2	Month 2	22-Feb-14	52	SD	
4	01-701-1015	OALLRESP	Overall Response	3	Month 3	24-Mar-14	82	PR	Y
5	01-701-1015	OALLRESP	Overall Response	4	Month 4	22-Apr-14	111	PR	
6	01-701-1015	OALLRESP	Overall Response	5	Month 5	23-May-14	142	PR	
7	01-701-1015	OALLRESP	Overall Response	6	Month 6	18-Jun-14	168	PR	

6.3 Measurement of Longest Diameter Over Time Per Patient

Although the sum of the longest diameter is used to assess the change in tumor size, it can be beneficial to see how each individual tumor size is changing over time along with the change in the sum of the longest diameters. Output 6-1 is a graphical representation of these measurements over time. The longest diameter for each individual tumor is displayed using a series plot on the Y axis, while the sum of the longest diameters is displayed as bar charts on the Y2 axis. With the graph you can see how each individual tumor contributed to the sum. For example, at Screening, Lung had the largest contribution to the sum of all tumors, while Breast and Other were pretty close in size. However, as the study went on, Lung and Other contributed about the same to the sum of the diameters, while Breast tumors did not have as much contribution.

Output 6-1: Longest Diameter Measurements Over Time

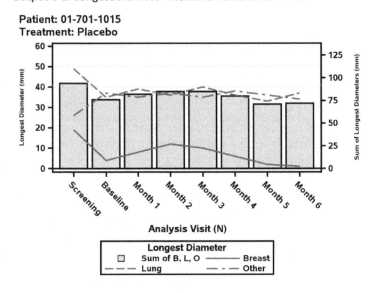

At first glance, producing this graph might seem very straightforward. With a little data manipulation of the data in Display 6-1, we could have the SUMLDIAM on the first record for each patient and visit.

Display 6-3: Data Snippet of Modified Analysis Data Set for Tumor Diameter (ADTMDMOD)

Obs	USUBJID	PARAMCD	TULOC	AVISIT	ADT	AVAL	SUMLDIAM
1	01-701-1015	LDIAM	Breast	Screening	3-Oct-13	19	94
2	01-701-1015	LDIAM	Lung	Screening	3-Oct-13	49	
3	01-701-1015	LDIAM	Other	Screening	3-Oct-13	26	
4	01-701-1015	LDIAM	Breast	Baseline	26-Dec-13	4	76
5	01-701-1015	LDIAM	Lung	Baseline	26-Dec-13	35	
6	01-701-1015	LDIAM	Other	Baseline	26-Dec-13	37	
7	01-701-1015	LDIAM	Breast	Month 1	25-Jan-14	8	82
8	01-701-1015	LDIAM	Lung	Month 1	25-Jan-14	39	
9	01-701-1015	LDIAM	Other	Month 1	25-Jan-14	35	

With the data in a desirable format as shown in Display 6-3, your first inclination might be to use PROC SGPLOT as shown in Program 6-1 to produce the graph in Output 6-1.

Program 6-1: Bar Chart with Series Plot with PROC SGPLOT

```
proc sgplot data = adam.adtmdmod;
   by trtp usubjid;

   vbar avisitn / response = sumldiam y2axis;
   series x = avisitn y = aval / group = tuloc;

   xaxis type = discrete label = 'Analysis Visit';
   yaxis type = linear label = 'Longest Diameter (mm)';
   y2axis type = linear label = 'Sum of Longest Diameters (mm)';
run;
```

Upon execution of Program 6-1, it generates the following error message:

```
ERROR: Attempting to overlay incompatible plot or chart types.
```

This is because within PROC SGPLOT there are certain plots that are incompatible. For example, series plots are not compatible with histograms and bar charts (Wicklin, 2016). Thus, using PROC SGPLOT it is not possible to produce a graph that has both a bar chart and series plot without first summarizing the data that is used in the bar chart. If the data is summarized previously, then you can use PROC SGPLOT, but you would need to use VBARPARM statement instead of VBAR since VBARPARM is compatible with SERIES. However, if the data you have is not yet summarized as is the case with the data in Display 6-3, then in order to produce a graph with a bar chart and series plot as illustrated in Output 6-1, you will need to turn to GTL for the generation of the desired output as demonstrated in Program 6-2.

Program 6-2: Bar Chart with Series Plot Using GTL

```
proc format;
   picture vis ❶
            -1 = 'Screening' (noedit)
             0 = 'Baseline' (noedit)
       1 - high = 09 (prefix = 'Month ');
run;

proc template;
   define statgraph tumd;
      begingraph;
         dynamic _byval_ _byval2_;
         entrytitle halign = left textattrs = (size = 11pt)
             "Patient: " _byval2_;
         entrytitle halign = left textattrs = (size = 11pt)
             "Treatment: " _byval_;

         layout overlay / xaxisopts = (type = linear
                              linearopts = (tickvaluesequence = ❷
                                                (start = -1 end = 6
                                                 increment = 1)
                                            tickvaluefitpolicy=rotatealways ❸
                                            tickvaluerotation = diagonal)) ❸
                          yaxisopts = (type = linear
                                       label = 'Longest Diameter (mm)'
                                       labelattrs = (size = 7pt)
                                       linearopts = (viewmin = 0  ❹
                                                     viewmax = 60) ❹
                                       griddisplay = on  ❺
                                       gridattrs = (pattern = 35  ❺
                                                    color = libgr))
                          y2axisopts = (type = linear
                                        label = 'Sum of Longest Diameters (mm)'
                                        labelattrs = (size = 7pt)
                                        linearopts = (viewmin = 0  ❹
                                                      viewmax = 135));  ❹

            barchart x = avisitn y = sumldiam / yaxis = y2  ❻
                                                name = 'SumLD';  ❼
            seriesplot x = avisitn y = aval / group = tuloc ❽
                                              name = 'LDiam';  ❼

            discretelegend 'SumLD' 'LDiam'/ location=outside  ❾
                                           title = 'Longest Diameter';  ❾
         endlayout;
      endgraph;
   end;
run;

proc sgrender data = adam.adtmdmod template = tumd;
   by trtp usubjid;
   format avisitn vis.;
   where usubjid ? '1015';  ❿
run;
```

❶ The FORMAT procedure can be used to create a picture format to display the data. With the use of the PICTURE statement, there are several options that can be used:

- NOEDIT indicates that there is a preset value and does not require editing. Thus, for this data, any record with AVISITN = -1 will be displayed as "Screening" with no additional editing done to the display value. The same is true with AVISITN = 0. It will be displayed as "Baseline" with no additional editing.

- PREFIX indicates a label that precedes the variable value.

- The number to the right of the equal sign on the last row of the format, that is, "1 – high = 9 (prefix = 'Month')" is used as a digit selector. If a 0 is used as a digit selector, then it is a place holder, and if there is a value, then it should be populated, but it is not required to be populated. If the digit selector is greater than 0, then it is a place holder that must be populated. Therefore, in this case, any AVISITN value greater than 1 will be displayed as "Month #". For example, if AVISITN = 3, then it would be displayed as "Month 3"; if AVISITN = 12, then it would be displayed as "Month 12".

❷ For linear axes, you can indicate the start and end of the tick values as well as the increment value, by using the TICKVALUESEQUENCE option and specifying START (first tick mark), END (last tick mark) and INCREMENT. If TICKVALUESEQUENCE is not specified, then SAS determines the tick values based on the actual data.

❸ Similar to the FITPOLICY option in the XAXIS statement in PROC SGPLOT, the TICKVALUEFITPOLICY option specifies the policy to use when there is a possibility for a collision of the tick marks. ROTATEALWAYS indicates that the tick values should always be rotated regardless if there is a collision. TICKVALUEROTATION determines the degree of rotation:

- DIAGONAL represents a rotation of 45 degrees.

- DIAGONAL2 represents a rotation of -45 degrees.

- VERTICAL represents a rotation of 90 degrees.

❹ Additional options that you can use on linear axes are VIEWMIN and VIEWMAX. You can use these options with or without the TICKVALUESEQUENCE or TICKVALUELIST options. The VIEWMIN option indicates the smallest data point that should be displayed on the output and VIEWMAX option indicates the largest data point that should be displayed. If you do not indicate what the tick mark values should be on the axis, SAS determines what the values should be based on the specified VIEWMIN and VIEWMAX. So while the maximum value that can be displayed is 135, the Y2 axis only goes to 125, where 125 is determined based on an algorithm that is defined within SAS.

❺ Gridlines can help see where a point falls with respect to a particular value on the axis. To display the gridlines, you need to set GRIDDISPLAY = ON. In addition, you can control the various attributes associated with the gridlines, such as color and line pattern, by using the GRIDATTRS option.

❻ Because both the Y and Y2 axis are used, if you want to ensure that a graph uses the Y2 axis, you need to use the AXIS = Y2 option.

❼ The NAME option on a plot statement enables you to provide a name to the plot statement so that it can be referenced in another template statement, such as DISCRETELEGEND. The NAME option is primarily used when creating legends. For example, NAME = "SumLD" in the BARCHART statement sets up a reference used by the DISCRETELEGEND statement so that a legend entry is produced for the bar chart. The NAME = "LDiam" in the SERIESPLOT statement creates a reference also used by the DISCRETELEGEND statement so that legend entries for each tumor location is produced.

❽ A series plot can be generated in a similar fashion using GTL as it is done with SG procedures. In SG procedures the statement name is SERIES, but for GTL the statement to use is SERIESPLOT. Both use similar options such as grouping the data for each unique group value with the use of the GROUP option.

❾ DISCRETELEGEND references the other template plot statements using the NAME that each statement was assigned. This statement creates a legend for the graph. You can use different options to control the appearance of the legend. You can provide a title for the legend with the TITLE option. You can also control whether the legend appears inside or outside the plot area.

❿ For illustration purposes, only one patient is selected. However, this can be run on all patients by removing the WHERE clause. The BY statement includes USUBJID to allow for the printing of the patient ID on each individual graph.

6.4 Waterfall Plot of Best Overall Response with Best Percentage Change in Tumor Burden

One of the primary assessments in an oncology study is the tumor response to the study drug. How did the tumor burden change over time and what was the overall response when assessing the tumor burden against the standard criteria? If the best percentage change, the smallest PCHG value, is below the baseline, a negative value, then it is an indication of tumor shrinkage. If the best percentage change is above the baseline, a positive value, this is an indication of tumor growth. It is common to assume that shrinkage would correspond to a best overall response of complete or partial response (CR or PR) and that growth would correspond to progressive disease (PD). However, that is not always the case. With a waterfall plot, you can plot both the best overall response and the best percentage change on the same plot for each patient to see how they relate.

In order to produce a waterfall plot it is necessary to have both the best percentage change and best overall response for each patient on the same record. Thus, the data from Display 6-1 can be subsetted for only records where ANL01FL = "Y" as shown in Display 6-4. Recall that ANL01FL in Display 6-1 selects the best percentage change, that is, selecting the record with the smallest PCHG value for the patient across visits. In order to produce the waterfall plot, the data is sorted by treatment (TRTP) and decreasing values of percentage change (PCHG).

Display 6-4: Data Snippet of Analysis Data Set of Assessment for Percentage Change in Tumor Burden and Tumor Response (ADBORPCH)

Obs	SID	USUBJID	TRTP	PCHG	BOR
1	1	01-710-1077	Placebo	1.492537	SD
2	2	01-701-1047	Placebo	-1.923077	PR
3	3	01-703-1096	Placebo	-2.325581	SD
4	4	01-701-1015	Placebo	-6.578947	PR
5	5	01-706-1041	Placebo	-7.317073	SD
6	6	01-710-1060	Placebo	-7.692308	SD
7	7	01-701-1023	Placebo	-8.510638	NE
8	8	01-714-1035	Placebo	-11.76471	PR
9	9	01-716-1024	Placebo	-16.04938	CR
10	10	01-705-1059	Placebo	-17.39130	SD

When creating a waterfall plot for an oncology study, the first inclination might be to use the WATERFALL statement in PROC SGPLOT. However, this does not yield the type of waterfall plot needed. The WATERFALL statement accumulates the response based on the order of the data. We do not want to accumulate the data. Each patient should be represented by its own bar chart, as shown in Output 6-2. Program 6-3 illustrates how you can generate a waterfall plot to depict the needs for an oncology study.

Program 6-3: Waterfall Plots Using One VBAR Statement with GROUP Option

```
title1 Treatment:  #byval1;    ❶

proc sgplot data = adam.adborpch;
   by trtp;
   xaxis type = discrete display = none;
   yaxis type = linear minor
         label = 'Sum of Longest Diameters-Percentage Change from Baseline'
         grid gridattrs = (pattern = 35 color = libgr)    ❷
         minorgrid minorgridattrs = (pattern = 35 color = libgr);    ❷

   vbar sid / response = pchg ❸
              group = bor
              name = 'resp'
              barwidth = 0.7 ❹
              fillpattern;   ❺
run;
```

❶ If you want the title to contain information from the BY variables, the use of #BYVAL option uses the value from the corresponding BY variable. The number that follows #BYVAL indicates which BY variable should be used. Using #BYVAL1 displays the information from the first by variable.

❷ Program 6-2 demonstrated the use of gridlines for GTL. You can also use gridlines in SG procedures. Like with most statements and options in SG procedures, they have a slightly different name than their counterpart in GTL. For SG procedures, you use GRID and GRIDATTRS options to add gridlines and to control the appearance of the gridlines. In addition to the gridlines, you can indicate whether you want the minor gridlines to be displayed with the use of MINORGRID. The MINORGRIDATTRS option is used to control the attributes for the minor gridlines. Gridlines correspond to the major tick marks on the axis, while the minor gridlines correspond to the minor tick marks. For the gridlines, the color libgr corresponds to light_bluish_gray and has a HEX color code of CXB3B2BF. Colors can be referenced using one of seven color-naming schemes (Watson, 2020).

❸ In GTL, the bar chart is specified using the BARCHART plot statement and the ORIENT option to indicate whether the bar chart should be VERTICAL or HORIZONTAL. In SG procedures, HBAR and VBAR are used to indicate horizontal bar charts and vertical bar charts respectively. For both HBAR and VBAR, a category variable is required. RESPONSE represents the numeric response variable that is to be used in the plot.

❹ BARWIDTH indicates the width of each bar as a ratio of the maximum possible width, which is determined by finding the distance between the center of one bar to the center of the adjacent bars. The default value for BARWIDTH is 0.8.

❺ By default, when GROUP is specified, color rotation is used to distinguish between the different groups. However, if you want to use line fill pattern, such as diagonal lines or crosshatch, you need to use the FILLPATTERN option.

Output 6-2: Waterfall Plots Using One VBAR Statement with GROUP Option

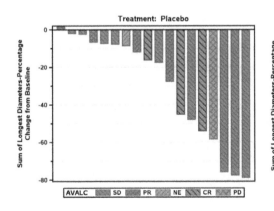
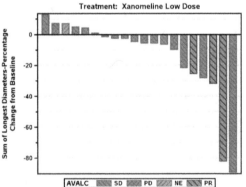

Using the approach shown in Program 6-3, notice that there is not a consistency in regards to how the bars for best overall response are represented. For Placebo, PR has a fill color of red with a fill pattern of crosshatched lines with light shading. But for Xanomeline Low Dose, PR has a fill color of brown with a fill pattern of right-slanting lines with a heavy shading. Having inconsistency from one graph to the next is not desirable.

An alternate approach to producing the waterfall plot using one VBAR statement with the GROUP option as shown in Program 6-3 is to treat each response value individually instead of using a GROUP option. In order to use the different approach the data in Display 6-4 needs to be modified so that each response value is indeed in its own variable, as shown in Display 6-5. Rather than have a separate variable for PCHG and BOR, the individual values for BOR become the variable names, then for each patient the PCHG is used to populate the corresponding BOR-value variable. For example, in Display 6-1 you see that patient 01-701-1015 had BOR = "PR" and PCHG = -6.578947. In the modified data set in Display 6-5, the PCHG is captured in the PR variable. In addition, the variable SID is added to help maintain a sequential sort order of the records based on the descending values of the PCHG within each TRTP.

Using the data in Display 6-5, Program 6-4 produces waterfall plots that have consistent fill colors and fill patterns for the best overall response.

Display 6-5: Data Snippet of Analysis Data Set of Assessment for Percentage Change in Tumor Burden and Tumor Response (ADBORPC2)

Obs	SID	USUBJID	TRTP	BOR	CR	PR	SD	PD	NE
1	1	01-710-1077	Placebo	SD			1.492537		
2	2	01-701-1047	Placebo	PR		-1.923077			
3	3	01-703-1096	Placebo	SD			-2.325581		
4	4	01-701-1015	Placebo	PR		-6.578947			
5	5	01-706-1041	Placebo	SD			-7.317073		
6	6	01-710-1060	Placebo	SD			-7.692308		
7	7	01-701-1023	Placebo	NE					-8.510638
8	8	01-714-1035	Placebo	PR		-11.76471			
9	9	01-716-1024	Placebo	CR	-16.04938				
10	10	01-705-1059	Placebo	SD			-17.39130		

Program 6-4: Waterfall Plots Using Individual VBAR Statements for Each Response Value

```
title1 Treatment:   #byval1;

proc sgplot data = adam.adborpc2;
   by trtp;
   xaxis type = discrete display = none;
   yaxis type = linear minor
          label = 'Sum of Longest Diameters-Percentage Change from Baseline'
          grid gridattrs = (pattern = 35 color = libgr)
          minorgrid minorgridattrs = (pattern = 35 color = libgr);

   vbar sid / response = cr name = 'CR' legendlabel = 'CR'
              barwidth = 0.7
              fillattrs = (color = cx6F7EB3)    ❶
              fillpattern fillpatternattrs = (color = cx445694    ❶
                                               pattern = L1)
              outlineattrs = (color = cx445694);  ❶
   vbar sid / response = pr name = 'PR' legendlabel = 'PR'
              barwidth = 0.7
              fillattrs = (color = cxD05B5B)
              fillpattern fillpatternattrs = (color = cxA23A2E
```

```
                                                      pattern = R1)
                      outlineattrs = (color = cxA23A2E);
    vbar sid / response = sd name = 'SD' legendlabel = 'SD'
                 barwidth = 0.7
                 fillattrs = (color = cx66A5A0)
                 fillpattern fillpatternattrs = (color = cx01665E
                                                 pattern = X1)
                 outlineattrs = (color = cx01665E);
    vbar sid / response = pd name = 'PD' legendlabel = 'PD'
                 barwidth = 0.7
                 fillattrs = (color = cx543005)
                 fillpattern fillpatternattrs = (color = cxA9865B
                                                 pattern = L4)
                 outlineattrs = (color = cxA9865B);
    vbar sid / response = ne name = 'NE' legendlabel = 'NE'
                 barwidth = 0.7
                 fillattrs = (color = cx9D3CDB)
                 fillpattern fillpatternattrs = (color = cxB689CD
                                                 pattern = R4)
                 outlineattrs = (color = cxB689CD);

    keylegend 'CR' 'PR' 'SD' 'PD' 'NE' / border title = 'Best Overall Response'; ❷
run;
```

❶ In order for each response to consistently have the same fill color, outline color, fill pattern, and fill pattern color, then each response needs to be its own plot statement with the appropriate options specified.

- FILLATTRS is used to specify the color and transparency for the fill area of the bar chart. Colors can be specified using one of seven color-naming schemes, which includes HEX codes. If you have the red, green, blue color values that you want to use, there are SAS color-utility macros that can be used to help convert to one of the allowed color-naming schemes (Watson, 2020).

- FILLPATTERNATTRS is used to specify the line pattern and the color of the line pattern for the fill pattern of the bar chart. In order for the fill patterns to be displayed, the FILLPATTERN option must also be specified. There are 15 patterns available. Each pattern is represented by two characters. The first character indicates the line direction (R for right, L for left, X for crosshatch). The second represents the type of line. Values 1–3 represent a single weight and is the weight of the line. Values 4–5 are used for double lines, with 4 being a thin weight and 5 being a medium thickness.

- OUTLINEATTRS is used to specify the attributes for the outline of the bar chart.

❷ KEYLEGEND adds a legend to the plot using the referenced plot names. The BORDER option indicates that a border should be drawn around the legend. The TITLE option enables you to specify a title for the legend.

Notice that in Program 6-3 there is no legend statement. This is because by default a legend will be generated for some plots within SGPLOT.

Output 6-3: Waterfall Plots Using Individual VBAR Statements for Each Response Value

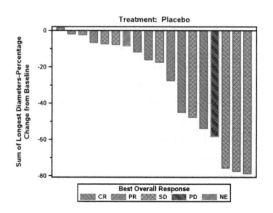
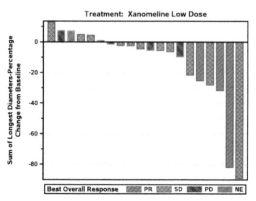

6.5 Spider Plot of Percentage Change from Baseline for Each Patient Over Time

Section 6.3 *Measurement of Longest Diameter Over Time Per Patient* illustrated an individual patient's data, but what if you want to look at the percentage change of all patients on the same output? A spider plot enables you to visualize the tumor response for all patients over time. With a spider plot you, have a reference point, such as baseline visit, and determine the days from that reference point to the visit.

Display 6-6 shows the data used to produce the spider plot. Notice that a new study day variable (A_BASEDY) is introduced, which is based on the baseline date (BASEDT). Program 6-5 uses data in Display 6-6 to produce the spider plot shown in Output 6-4.

Display 6-6: Data Snippet of Analysis Data Set for Percentage Change in Tumor Burden and Tumor Response (ADTMBSDY)

Obs	USUBJID	TRTP	ADT	BASEDT	A_BASEDY	PCHG
1	01-701-1015	Placebo	3-Oct-13	26-Dec-13	-84	23.68421
2	01-701-1015	Placebo	26-Dec-13	26-Dec-13	0	0
3	01-701-1015	Placebo	25-Jan-14	26-Dec-13	31	7.894737
4	01-701-1015	Placebo	22-Feb-14	26-Dec-13	59	11.84211
5	01-701-1015	Placebo	24-Mar-14	26-Dec-13	89	11.84211
6	01-701-1015	Placebo	22-Apr-14	26-Dec-13	118	5.263158
7	01-701-1015	Placebo	23-May-14	26-Dec-13	149	-6.57895
8	01-701-1015	Placebo	18-Jun-14	26-Dec-13	175	-5.26316
9	01-701-1023	Placebo	2-May-12	22-Jul-12	-81	-85.1064
10	01-701-1023	Placebo	22-Jul-12	22-Jul-12	0	0

Program 6-5: Spider Plot of Percentage Change from Baseline for Each Patient Over Time

```
title1 Treatment:   #byval1;

proc sgplot data = adam.adtmbsdy noautolegend;  ❶
   by trtp;
   xaxis type = linear label = 'Days from Baseline Visit'
         values = (-100 to 180 by 40)   ❷
         grid gridattrs = (pattern = 35 color = libgr)
         minorgrid minorgridattrs = (pattern = 35 color = libgr);
   yaxis type = linear minor
         label = 'Sum of Longest Diameters-Percentage Change from Baseline'
         grid gridattrs = (pattern = 35 color = libgr)
         minorgrid minorgridattrs = (pattern = 35 color = libgr);

   series x = a_basedy y = pchg / group = usubjid;
   refline 0 / axis = X;   ❸
   refline 1 / axis = Y;   ❸
run;
```

❶   If you want to suppress the legend from automatically being generated for the plots within the SG procedure, you need to use the NOAUTOLEGEND option.

❷   The VALUES option on the axis statement is equivalent to the TICKVALUESEQUENCE and TICKVALUELIST that is used in GTL. However, the specification is slightly different. For example, in GTL, it would be specified as TICKVALUESEQUENCE = (START = -100 END = 180 INCREMENT = 40). With the VALUES statement you can provide a list of values either in ascending or descending order (for example, 10 20 30). Or you can provide a range that would be incremented by a specific unit as illustrated in Program 6-5.

❸   To create a vertical line or a horizontal line use the REFLINE statement. AXIS option indicates whether you draw a vertical line (AXIS = X) or a horizontal line (AXIS = Y).

Output 6-4: Spider Plot of Percentage Change from Baseline for Each Patient Over Time

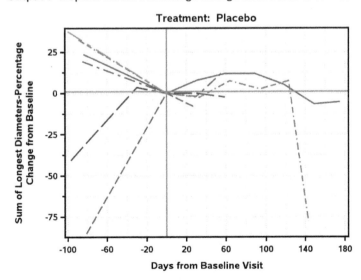

6.6 Swimmer Plot for Overall Response During Course of Study

Waterfall plots are ideal for showcasing the best overall response for each patient. However, if you are interested in assessing the tumor burden during the course of a study for each patient, then a swimmer plot is ideal.

Swimmer plots are discussed in Chapter 4, but here we will take a different approach to creating one. As with most things, there are many different ways to create a swimmer plot.

Like all the other examples shown, the data needs to be formatted in a specific way in order to produce the desired output. The day on which a patient starts treatment is referred to as Study Day 1; thus, by default all patients on treatment will have STRT set to 1. The treatment duration (TRTDUR) is also needed. In order to identify the baseline time point, the study day associated with the baseline needs to be captured as a separate variable (BASEDY) and the study day for post-baseline assessments related to the next visit (NEXTDY) also needs to be captured on the current record. The PAGE variable is only used to identify what page in the output the patients' data are on. The PAGE variable is to prevent putting all patients' data on the same graph and making it unreadable. Note that we also introduce SUBJID, which is only the last four characters of USUBJID. The purpose of this is so that the shorter version of the patient identifier can be used on the output. Program 6-6 uses data from Display 6-7 in order to produce the swimmer plot shown in Output 6-5.

Display 6-7: Data Snippet of Analysis Data Set for Percentage Change in Tumor Burden and Tumor Response (ADRSPMOD)

Obs	SID	USUBJID	SUBJID	TRTP	STRT	TRTDUR	AVALC	ADY	BASEDY	NEXTDY	PAGE
53	11	01-709-1001	1001	Placebo	1	183		.	-18	.	2
54	11	01-709-1001	1001	Placebo	1	183	SD	14	.	44	2
55	11	01-709-1001	1001	Placebo	1	183	SD	44	.	76	2
56	11	01-709-1001	1001	Placebo	1	183	PD	44	.	44	2
57	11	01-709-1001	1001	Placebo	1	183	PD	76	.	104	2
58	11	01-709-1001	1001	Placebo	1	183	SD	104	.	.	2
59	12	01-710-1027	1027	Placebo	1	183		.	-25	.	2
60	12	01-710-1027	1027	Placebo	1	183		5	.	35	2
61	12	01-710-1027	1027	Placebo	1	183	SD	35	.	64	2
62	12	01-710-1027	1027	Placebo	1	183	SD	64	.	93	2

Program 6-6: Swimmer Plot for Overall Response During Course of Study

```
dataattrmap (drop = i);
   id = 'swimmer';   ❶
   show = 'attrmap';
   array vl(5) $2 _temporary_ ('CR' 'PR' 'SD' 'PD' ' ');
   array lc(5) $20 _temporary_ ('cornflowerblue' 'firebrick'
                                'orange' 'green' 'ghostwhite');
   do i = 1 to 5;
      value = vl(i);
      linecolor = lc(i);
      output;
   end;
run;

proc template;
   define statgraph disdur;
      begingraph / border = false;
         dynamic _byval2_;
            entrytitle halign = left textattrs = (size = 11pt)
               "Treatment: " _byval2_;

            layout overlay / xaxisopts = (label = 'Study Day'
                                          linearopts = (minorticks = true))
                             yaxisopts = (label = 'Patient ID'
                                          display = (label))
                             walldisplay = none;

               innermargin / align = left ❷
                             separator = true  ❸
                             separatorattrs = (color = black   ❸
                                               thickness = 2px);
                  axistable y = sid value = subjid /   ❹
                             valueattrs = (size = 9pt weight = bold) ❹
                             display = (values); ❹
               endinnermargin;

               highlowplot y = sid low = strt high = trtdur /   ❺
                             type = bar ❻
                             fillattrs = (color = ghostwhite)
                             outlineattrs = (color = black)
                             name = 'Trt Dur';

               vectorplot x = nextdy y = sid xorigin = ady yorigin = sid /
                             group = avalc ❼
                             arrowheads = false  ❽
                             lineattrs = (thickness = 7px pattern = solid)
                             name = 'resp';

               scatterplot x = basedy y = sid /
                             markerattrs = (symbol = star color = red)
                             name = 'Baseline' legendlabel = 'Baseline';

               discretelegend 'resp'/ location = outside  ❾
                             halign = left valign = bottom
                             exclude = (' ') border = false;
               discretelegend 'Trt Dur' 'Baseline'/ location = outside   ❾
                             halign = right valign = bottom
                             border = false;
```

```
        endlayout;
      endgraph;
   end;
run;

proc sgrender data = adam.adrspmod template = disdur dattrmap =attrmap;  ❿
  dattrvar avalc = 'swimmer';  ❿
  by page trtp;
run;
```

❶ In order to maintain consistency in the appearance from output to output for specific values, you can create an attribute map data set that can be referenced by different plot statements. In order to create an attribute map data set, it needs to contain an ID. In addition, there are reserved variable names that attribute map data sets use. The data set needs to have a record for each possible formatted value and specify the attributes that you want to control for those values. For example, if you only need LINECOLOR then there is no need to specify the other attributes, although it does not hurt to have them in the data set for use in other plots (Watson, 2020). The data set can be saved to a permanent location so that it can be referenced later and reused across multiple SG procedures. To learn more about defining an attribute map, visit Defining a Discrete Attribute Map at SAS® 9.4 Graph Template Language: Reference, Fifth Edition documentation (SAS Institute Inc., 2020).

❷ The INNERMARGIN block sets aside a nested region within the layout. This area is reserved for use by a BLOCKPLOT or an AXISTABLE.

❸ If you want to show a clear distinction between the inner margin area and the rest of the plot, you can use the SEPARATOR = TRUE option. The SEPARATORATTRS option is used to handle the appearance of the separator line. By default the separator line is not displayed (that is SEPARATOR = FALSE).

❹ In Program 6-5, an AXISTABLE statement is used to display the patient ID so that each floating bar can be associated with the patient it pertains to. Similar to other GTL statements, you can control the appearance using the different options. For this graph, if you want to bold the IDs and display only the values of the AXISTABLE, then the VALUEATTRS option enables you to set the various attributes for the value, such as the SIZE and WEIGHT. The DISPLAY option indicates what you want to display. In this example, if you only want to display the values, in other words, you do not want the label associated with the variable to be displayed, then you can use DISPLAY = (VALUES) to indicate only the values should be displayed; otherwise, by default the values and the variable label will be displayed.

❺ In order to produce the black bar that spans the patients' treatment durations, you can use the HIGHLOWPLOT statement. The HIGHLOWPLOT produces either a vertical or horizontal line based on the LOW and HIGH value. The orientation is dependent on whether you use the X or Y option. X indicates a vertical line while Y indicates a horizontal line.

❻ Within the HIGHLOWPLOT, you can indicate whether you want a line or a bar with the TYPE option. If you specify TYPE = LINE, then you can control the appearance with the LINEATTRS option. If you specify TYPE = BAR, then you can control the appearance with the use of FILLATTRS and/or OUTLINEATTRS options as illustrated in Program 6-5. For example, in the swimmer plot, the bar used to represent the patients' treatment durations uses the color ghostwhite to fill the bar and black to produce the outline around the bar.

❼ To combine the patients' treatment durations with their responses, you can overlay the patients' responses using the VECTORPLOT. VECTORPLOT draws directional lines from the point of origin, which is denoted by XORIGIN and YORIGIN arguments. The endpoint for the directional line is represented by X and Y. The GROUP implements the color rotation based on the style that is used or based on an attribute map as is the case with this example. The GROUP option allows for a consistent color scheme for each unique value of AVALC.

❽ Unlike the HIGHLOWPLOT, which allows for end caps on both ends of the floating bar, the VECTORPLOT is directional and an end cap can only be displayed on the end, which is represented by X and Y. The end cap is either no end cap or an arrowhead. To indicate that no arrowhead (or end cap) is to be displayed, you need to use the ARROWHEADS = FALSE option. The default is to show OPEN arrowheads.

❾ DISCRETELEGEND statement creates a legend entry for each unique value for each plot name referenced. The legend typically consists of a symbol, such as a marker or a line or a marker and line combination, depending on the type of output displayed to represent that component on the plot. By default the location of the legend is placed outside the plot area and it is centered at the bottom. It is not necessary to specify LOCATION = OUTSIDE. It is done here to illustrate the use of the option. In some cases it might be fine to use the default location, but in the case where you might have more than one legend, you do not want a collision where the subsequent legends are overwriting the previous legends. To avoid this, you can use HALIGN and VALIGN to indicate what space that particular legend should occupy within the graph. If a specific value is not to be displayed in the legend, then the EXCLUDE option can be used to suppress those values as illustrated in the first DISCRETELEGEND statement in Program 6-5.

❿ In order to reference the attribute map that has been saved as a data set, you need to use the DATTRMAP option in the PROC SGRENDER statement. The attribute map can have multiple attributes maps defined within the data set, to associate the correct attribute map with the correct variable, the use of DATTRVAR statement is needed. You can reference multiple associations using the DATTRVAR statement.

Output 6-5: Swimmer Plot for Overall Response During Course of Study

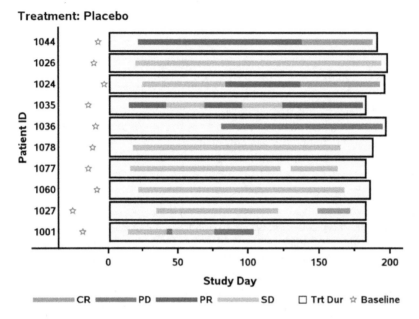

6.7 Other Types of Plots for Oncology Data

Two additional types of visualizations that are very common with oncology data are survival plots and forest plots. Survival plots are used to assess things such as progression-free survival, overall survival, as well as other time-to-event endpoints. In addition, survival plots can show the number of patients at risk for each time point and at which point there was censoring. Forest plots help with assessing the effect of covariates on an outcome.

Although these are outputs that are common in oncology data, they will not be discussed in this chapter since the concepts to generate them are similar to the concepts discussed in Chapter 5.

6.8 Chapter Summary

Oncology studies are inherently challenging. Understanding what the key endpoint is, tumor shrinkage, helps to grasp what types of graphs are needed. Due to the nature of oncology data, specific types of graphs are necessary in order to get a visual representation of how the tumor burden is responding to treatment. Regardless if the graph is for an individual patient over time, or if it is for all patients on one output, the end result is to determine the tumor burden response. With the visual representations, it can be more easily detected if the patient is responding to treatment based on the overall response and the percentage change in tumor burden.

References

Carpenter, A. L. (1998). "Better Titles: Using The #BYVAR and #BYVAL Title Options." Proceedings of the Twenty-Third Annual SAS Users Group International Conference. Cary, NC: SAS Institute Inc. Nashville: SUGI. Retrieved from https://support.sas.com/resources/papers/proceedings/proceedings/sugi23/Coders/p75.pdf

Eisenhauer, E. A., Therasse, P., Bogaerts, J., Schwartz, L. H., Sargent, D., Ford, R., . . . Verweij, J. (2009). "New response evaluation criteria in solid tumours: Revised RECIST guideline (version 1.1)." *European Journal of Cancer,* 45, 228-247. Retrieved from https://ctep.cancer.gov/protocolDevelopment/docs/recist_guideline.pdf

Matange, S. (2013). *Getting Started with the Graph Template Language in SAS®: Examples, Tips, and Techniques for Creating Custom Graphs*. Cary, NC: SAS Institute Inc.

Matange, S. (2019). "Effective Graphical Representation of Tumor Data." Proceedings of the Annual Pharmaceutical SAS Users Group Conference. Cary, NC: SAS Institute Inc. Philadelphia: PharmaSUG.

National Cancer Institute. "What is Cancer?" About Cancer: Understanding Cancer. (2015, February 9). Retrieved 2020, from National Cancer Institute: https://www.cancer.gov/about-cancer/understanding/what-is-cancer

NCI Dictionaries. (n.d.). "Recist." Retrieved from National Cancer Institute: https://www.cancer.gov/publications/dictionaries/cancer-terms/def/recist

Pandey, M. (2018). "Unleashing Potential of Graphs for Oncology Trials." Proceedings of the Pharmaceutical Users Software Exchange 2018 Conference. Raleigh: PHUSE.

Ruel, N. H. and& Frankel, P. H. (2016). "Graphical Results in Clinical Oncology Studies." Proceedings of the SAS Global Forum 2016 Conference. Las Vegas: Cary, NC: SAS Institute Inc. Global Forum.

SAS Institute Inc. (2020, July 30). SAS Graph Template Language: Reference: Defining a Discrete Attribute Map. Retrieved from SAS® 9.4 and SAS® Viya® 3.5 Programming Documentation. *SAS® 9.4 Graph Template Language: Reference, Fifth Edition*: https://documentation.sas.com/?docsetId=grstatgraph&docsetTarget=n1oq0axfb1tc1yn10kgw6dyuitk2.htm&docsetVersion=9.4&locale=en

Watson, R. (2019). "Great Time to Learn GTL: A Step-by-Step Approach to Creating the Impossible." Proceedings of the SAS Global Forum 2019 Conference. SAS Institute Inc. Dallas: SAS.

Watson, R. (2020). "What's Your Favorite Color? Controlling the Appearance of a Graph." Proceedings of the SAS Global Forum 2020 Conference. Cary, NC: SAS Institute Inc.

Wicklin, R. (2016, September 12). "Overlay a curve on a bar chart in SAS." SAS Blogs: The DO Loop. Retrieved from SAS: https://blogs.sas.com/content/iml/2016/09/12/overlay-curve-bar-chart-sas.html

Chapter 7: Using SAS to Generate Special Purpose Graphs

7.1 Introduction to Special Purpose Graphs

In this chapter, you will learn how to create nonstandard clinical graphs, such as interactive graphs and Venn diagrams. These types of graphs are seldom requested; however, they are fun to produce and it is worth knowing how to produce them.

Using interactive graphs can help you interrogate the data further and provide you with a better understanding of the data because you are able to assess the relationships between the outputs that you are investigating.

Likewise, Venn diagrams can help you evaluate the association of elements between different groups. For example, with Venn diagrams, you can see the adverse events in common between two treatment groups, and the adverse events that are mutually exclusive. That is, the events that do not occur in both treatment groups.

7.2 Interactive Graphs

This section demonstrates how you can use interactive graphics to assess and report your AE data. You will also see how to display "details-on-demand" when you hover over a point.

Adding interactivity to your graphs will enliven your data. Having the ability to drill down on your graphs enables you to interrogate your graphs and understand your data better.

Creating drill-down graphs requires the use of HTML. In addition to understanding the basics of creating a graph as illustrated in previous chapters, you need to know a few supplemental things, such as using:

- ODS OUTPUT HTML statements around your SG procedures or your GTL statements

- URL = *<CHARACTER-VARIABLE>* option in your plot statements

- IMAGEMAP = ON option within ODS GRAPHICS

You can also create graphs that can be filtered when you select items from a drop-down menu. You can achieve this by using the same drill-down methods that were previously mentioned, along with using a few additional lines of HTML code to create your drop-down menu. For you to be able to filter on selected items from a drop-down menu, you also will need to use JavaScript and Cascading Style Sheets (CSS). The JavaScript and CSS code used in this chapter are made available at https://github.com/rwatson724/SAS-Graphs-Clinical-Trials-Example for your use. Therefore, it is not necessary to understand how to generate the JavaScript and CSS code.

The adverse event data set, *Analysis Dataset for Adverse Events* (ADAE), used in this section is from the CDISC SDTM/ADaM Pilot Project.

7.2.1 Illustrating Adverse Event Drill Down

To view an interactive graph with a drill down affect for Adverse Events (AE) plots, you can visit https://github.com/rwatson724/SAS-Graphs-Clinical-Trials-Example.

When incorporating a drill-down feature, it is essential to know what you want to display after you have drilled down. For example, when you hover over the graph, do you want to display the active link that can take you to another graph or a table, or display basic information about that specific item in the graph? In the interactive AE example found on the GitHub website, https://github.com/rwatson724/SAS-Graphs-Clinical-Trials-Example, the main graph displays the percentage of patients with at least one AE in each treatment group for each AE. When you hover over one of the bars on the main graph, you see the actual percentage of patients for that AE and treatment group combination as seen in Output 7-1.

Output 7-1: Main Graph – Hover Over Bar

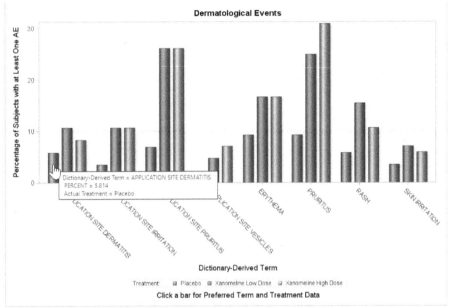

If you click on the bar as highlighted in Output 7-2, you will be taken to a secondary graph, which shows the number of patients who had the AE by each severity group, such as Output 7-3. Output 7-3 shows the number of patients who had the Application Site Dermatitis AE by each severity group.

Output 7-2: Main Graph – Clicking on Bar

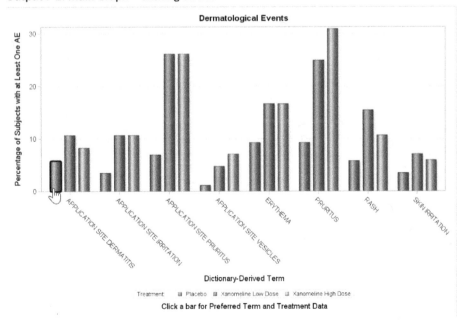

Output 7-3: Example Secondary Graph

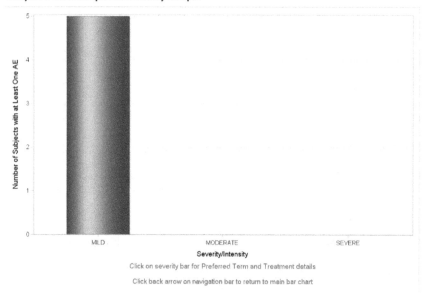

If you want to see the patients that had that AE and severity, then you would click on the bar that corresponds to the AE and severity of interest. Output 7-4 shows you how the graph will look just before you click on the mild bar of the patients that had Application Site Dermatitis. When you click on the bar, which is the second drill down, you are taken to a listing of the patients with that specific AE and severity, as seen in Output 7-5. Output 7-5 shows a listing of the placebo patients who had a mild Application Site Dermatitis event. You can see that five patients had a mild Application Site Dermatitis event.

Output 7-4: Example Secondary Graph – Clicking on Bar

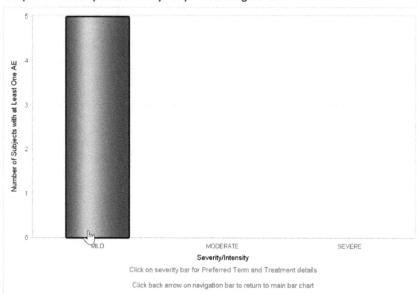

Output 7-5: Patient Listing

Click back arrow on navigation bar to return to severity bar chart								
Body System or Organ Class=GENERAL DISORDERS AND ADMINISTRATION SITE CONDITIONS Dictionary-Derived Term=APPLICATION SITE DERMATITIS Severity/Intensity=MILD Actual Treatment=Placebo								
Unique Subject Identifier	Reported Term for the Adverse Event	Analysis Start Date	Analysis End Date	AE Duration (N)	Serious Event	Causality	AREL	Outcome of Adverse Event
01-709-1088	APPLICATION SITE DERMATITIS	08MAY2014			N	PROBABLE	RELATED	NOT RECOVERED/NOT RESOLVED
01-709-1312	APPLICATION SITE DERMATITIS	17MAY2014		.	N	PROBABLE	RELATED	NOT RECOVERED/NOT RESOLVED
01-710-1077	APPLICATION SITE DERMATITIS	12DEC2013	12DEC2013	1	N	POSSIBLE	RELATED	RECOVERED/RESOLVED
01-713-1256	APPLICATION SITE DERMATITIS	30OCT2012	31OCT2012	2	N	PROBABLE	RELATED	NOT RECOVERED/NOT RESOLVED
	APPLICATION SITE DERMATITIS	30OCT2012	31OCT2012	2	N	PROBABLE	RELATED	RECOVERED/RESOLVED
01-718-1355	APPLICATION SITE DERMATITIS	27APR2013	27APR2013	1	N	PROBABLE	RELATED	RECOVERED/RESOLVED
	APPLICATION SITE DERMATITIS	12MAY2013	12MAY2013	1	N	PROBABLE	RELATED	RECOVERED/RESOLVED
	APPLICATION SITE DERMATITIS	16MAY2013	16MAY2013	1	N	PROBABLE	RELATED	RECOVERED/RESOLVED
	APPLICATION SITE DERMATITIS	20MAY2013	20MAY2013	1	N	PROBABLE	RELATED	RECOVERED/RESOLVED

When building a drill-down graph, you generally work backward, since the file you need to link to needs to already exist. Using the AE example, drill-down capability was achieved by:

1. Producing patient-level listings of each AE, for each treatment, and severity group.
2. Producing graph of counts of AEs, for each treatment, and severity group.
   a. Linking these bars to the patient level listings in step 1.
3. Producing a graph of the percentage of patients with AEs by each treatment group.
   a. Linking these bars to the counts of AEs in step 2.

This section focuses on creating the interactivity between the graphs, which show the percentage of patients with the AEs by each treatment group and the graphs that show the counts of AEs, for each treatment and severity group combination. If you would like to see more details about also producing the patient level listings of each AE, for each treatment, and severity group, then refer to Harris and Watson (2019).

7.2.1.1 Producing Graphs of Adverse Event Counts for each Treatment and Severity Group

The data to produce the AE counts, for each treatment and severity group is shown in Display 7-1 and is a data snippet from the ADAESEV data set. The ADAESEV data set uses the ADAE data set to calculate the number of patients that had at least one AE in each AE and severity group. The following variables in Display 7-1 are:

- TRTA: the actual treatments;
- AEDECOD: the adverse events;
- AESEV: the severity of the AE;
- PREFTERM: the AE name in lowercase text;
- COUNT: the results of the number of patients that had at least one AE for that particular AE, treatment, and severity level.

Display 7-1: Data Snippet of Number of Patients with at Least One Adverse Event by Severity (ADAESEV)

Obs	TRTA	AEDECOD	AESEV	PREFTERM	COUNT
1	Placebo	APPLICATION SITE DERMATITIS	MILD	application_site_dermatitis	5
2	Placebo	APPLICATION SITE DERMATITIS	MODERATE	application_site_dermatitis	0
3	Placebo	APPLICATION SITE DERMATITIS	SEVERE	application_site_dermatitis	0
4	Placebo	APPLICATION SITE IRRITATION	MILD	application_site_irritation	2
5	Placebo	APPLICATION SITE IRRITATION	MODERATE	application_site_irritation	2
6	Placebo	APPLICATION SITE IRRITATION	SEVERE	application_site_irritation	0
7	Placebo	APPLICATION SITE PRURITUS	MILD	application_site_pruritus	6
8	Placebo	APPLICATION SITE PRURITUS	MODERATE	application_site_pruritus	1
9	Placebo	APPLICATION SITE PRURITUS	SEVERE	application_site_pruritus	0
10	Placebo	APPLICATION SITE VESICLES	MILD	application_site_vesicles	1
11	Placebo	APPLICATION SITE VESICLES	MODERATE	application_site_vesicles	0
12	Placebo	APPLICATION SITE VESICLES	SEVERE	application_site_vesicles	0
13	Placebo	ERYTHEMA	MILD	erythema	6
14	Placebo	ERYTHEMA	MODERATE	erythema	4

Program 7-1 shows how to produce interactive graphs of the AE counts for each AE, treatment, and severity group. Program 7-1 contains a macro that processes each treatment individually so that they can have a set color. The macro's purpose is to ensure that each treatment has a consistent color in the various AE graphs and assigning the graph output file names accordingly, such as making the graphs identifiable by their AE and the treatment group.

Program 7-1: AE Counts, for Each AE, Treatment, and Severity Group

```
   %macro genchart(trt = , color =);
    /* create a template for the supporting AE graph - within in macro so bar
colors
      can align with main graph */
     proc template;
        define statgraph sevfreq&trt;
           begingraph;
              entryfootnote textattrs = (size = 8 pt color = green)
                  "<Text1>"; ❶
              entryfootnote " ";
              entryfootnote textattrs = (size = 8 pt color = red)
                  "<Text2>"; ❶
              layout overlay / yaxisopts = (label = "<Label1>"
                                            griddisplay = on
                                            gridattrs = (color = lightgray
                                           pattern = dot));
                 barchart category = aesev response = count /
                                           barwidth = 0.5
                                           fillattrs = (color = &color)❷
                                           dataskin = sheen
                                           url = url1
                                            tip = (count);
              endlayout;
           endgraph;
        end;
     run;

     %let y = 1;
     %let upt = %scan(&allprefterm, &y, '#');   ❸

     %do %while (&upt ne   );

        %if %sysfunc(find(&upt, &trt)) > 0 %then %do;

           /* Generate the graph using ODS HTML. */
           ods html path = outp file = "&upt..html";   ❹
           proc sgrender data = adam.adaesev template = sevfreq&trt;
              where pttrt = "&upt";   ❺
              by aedecod trta;
           run;
           ods html close;
        %end;

        %let y = %eval(&y + 1);
        %let upt = %scan(&allprefterm, &y, '#');
     %end;
   %mend genchart;

   options mprint mlogic;

   %genchart(trt = 0,  color = cornflowerblue)
   %genchart(trt = 54, color = indianred)
   %genchart(trt = 81, color = cadetblue)
```

❶ This GTL code creates the bar charts. Due to space limitations, some of the labels and options have been omitted, and in some instances, this is denoted by <Text1> and <Text2>. However, the full SAS code with all the labels and options are in the SAS programs, which are available for you to download at https://github.com/rwatson724/SAS-Graphs-Clinical-Trials-Example.

❷ The COLOR = &COLOR option in the BARCHART statement assigns the bar chart the colors specified in the macro variable. This color assignment helps to ensure that each treatment has a set color.

❸ The *ALLPREFTERM* macro variable contains a string of AEs and treatment numbers separated by a #. A sample of the string is "application_site_dermatitis_0#application_site_dermatitis_54". When $Y = 1$, the macro variable *UPT* resolves to "application_site_dermatitis_0". There are eight dermatological AEs, and there are three treatments. Therefore, there are twenty-four unique values of AEs, for the AEs and treatment combination.

❹ ODS HTML statement creates the HTML file of the bar chart of AE counts by severity, for each of the 24 combinations of AE, and treatment group in this example. It uses the macro variable *UPT* to name the file that is generated.

❺ To be able to use interactive graphs, a graph has to be created for each active link in the preceding graph. Thus, counts of AEs by severity outputs are produced for each AE and treatment combination. In order for each treatment to have a consistent bar chart color across all AEs, a macro is used to associate the correct colors. Therefore, the WHERE statement is used to ensure the appropriate AE and treatment data is subsetted and the appropriate graph template, which utilizes the correct color for the treatment is used.

Output 7-6 illustrates an example of a graph showing the number of patients with at least one AE for each treatment and severity group. This particular graph shows the number of placebo patients for the Application Site Dermatitis AE. The use of the *SAS color name* cornflowerblue as a color for placebo in the macro call ensures that the template uses the *color* cornflower blue for all Severity/Intensity graphs for each AE in the placebo group.

Output 7-6: Number of Placebo Patients with at Least One Application Site Dermatitis by Severity

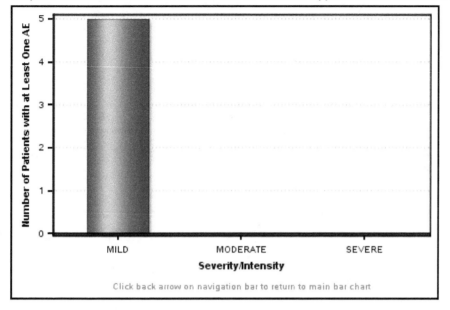

7.2.1.2 Producing Main Interactive Bar Chart

You have seen how to create the bar charts that the main graph will link to. Now you can turn your focus on how to create the main bar chart. The main bar chart shows the percentage of patients with at least one AE by the AE term, and grouped by treatment. When one of the bars is selected, it links to the bar charts created in Program 7-1, and a page showing the number of patients with at least one AE for that particular AE term, split by severity group, is shown (Output 7-6).

The data to produce the interactive bar chart is shown in Display 7-2. The following variables in Display 7-2 are:

- TRTA: the actual treatments;
- AEDECOD: the adverse events;
- URL0: the HTML pages that link from the main bar chart to the corresponding AE and treatment group bar chart. All of the twenty-four outputs produced by Program 7-1 have their file names comprised of the AE term and the numerical value of the treatment, as shown in the URL0 variable in Display 7-2. The use of URL0 is how the main graph will be able to link to the twenty-four outputs;
- COUNT: the results of the number of patients that had at least one AE;
- TOTCNT: the number of patients in each treatment group;
- PERCENT: the results of the percentage of patients that had at least one AE. This is derived from (COUNT / TOTCNT) * 100.

Display 7-2: Data Snippet of Percentages of Patients with at Least One Adverse Event (PTCNT2)

Obs	TRTA	AEDECOD	URL0	COUNT	TOTCNT	PERCENT
1	Placebo	APPLICATION SITE DERMATITIS	application_site_dermatitis_0.html	5	86	5.8140
2	Placebo	APPLICATION SITE IRRITATION	application_site_irritation_0.html	3	86	3.4884
3	Placebo	APPLICATION SITE PRURITUS	application_site_pruritus_0.html	6	86	6.9767
4	Placebo	APPLICATION SITE VESICLES	application_site_vesicles_0.html	1	86	1.1628
5	Placebo	ERYTHEMA	erythema_0.html	8	86	9.3023
6	Placebo	PRURITUS	pruritus_0.html	8	86	9.3023
7	Placebo	RASH	rash_0.html	5	86	5.8140
8	Placebo	SKIN IRRITATION	skin_irritation_0.html	3	86	3.4884
9	Xanomeline Low Dose	APPLICATION SITE DERMATITIS	application_site_dermatitis_54.html	9	84	10.7143
10	Xanomeline Low Dose	APPLICATION SITE IRRITATION	application_site_irritation_54.html	9	84	10.7143
11	Xanomeline Low Dose	APPLICATION SITE PRURITUS	application_site_pruritus_54.html	22	84	26.1905
12	Xanomeline Low Dose	APPLICATION SITE VESICLES	application_site_vesicles_54.html	4	84	4.7619

Program 7-2 uses the data shown in Display 7-2 to produce the main bar chart of the percentages of patients with at least one AE. The bar chart contains a grouping of percentages by treatment for each AE. This bar chart is considered to be the main graph because it will be the primary graph, and from this graph, you will be able to drill down to the supporting graphs that are produced by Program 7-1.

Program 7-2: Main Bar Chart of the Percentages of Patients with at Least One AE, by AE Term and Treatment Group

```
proc template;
   define statgraph ptfreq;
      begingraph;
         entrytitle textattrs = (size = 13pt) "Dermatological Events";
         entryfootnote textattrs = (size = 12pt)
            "Click a bar for Preferred Term and Treatment Data";
         layout overlay / yaxisopts = (label =
            "Percentage of Patients with at Least One AE"
                               labelattrs = (size=10pt)
                               tickvalueattrs = (size=10pt)
                               griddisplay = on
                               gridattrs = (color = lightgray
                                     pattern = dot))
```

```
                        xaxisopts = (labelattrs = (size=10pt)
                                     tickvalueattrs = (size=10pt));

            barchart category = aedecod response = percent /
                                       name = "dermevent"
                                       group = trta
                                       groupdisplay = cluster
                                       barwidth = 0.75
                                       dataskin = sheen
                                       url = url0
                                       tip = (url0);  ❶
            discretelegend "dermevent" / title = "Treatment:"
                                       valueattrs= (size=10pt);
        endlayout;
      endgraph;
    end;
run;

ods html path = outp file = "aes.html";

ods graphics / reset imagename = "aes" height = 8in width = 12in
               imagemap = on  ❷
               drilltarget = "_top";  ❸

proc sgrender data = tfldata.ptcnt2 template = ptfreq;  ❹
run;
ods html close;
```

❶ The variable URL0 is used in the URL option. This variable contains the values of the URL name associated with each AE and treatment group by severity bar charts, and when you click on the associated bar you are then navigated to the appropriate severity bar charts, which are created in Program 7-1.

❷ The IMAGEMAP = ON option helps to ensure that the links between the main bar chart and the AE counts by severity outputs work appropriately.

❸ The DRILLTARGET option specifies how the drill-down output is displayed in a window. There are five options to choose from. DRILLTARGET = "_top" opens the drill-down output in the full body of the window.

❹ PROC SGRENDER associates the PTCNT2 data set with the PTFREQ template to output the main bar chart.

Output 7-7 shows the bar charts of the percentages of patients with at least one AE, by AE, grouped by treatment. When you click on one of the bars, you will be linked to an associated page that shows the number of patients with at least one AE for that particular AE term, split by severity group (Output 7-6).

Output 7-7: Main Bar Chart of Percentages of Patients with at Least One AE, for Each AE

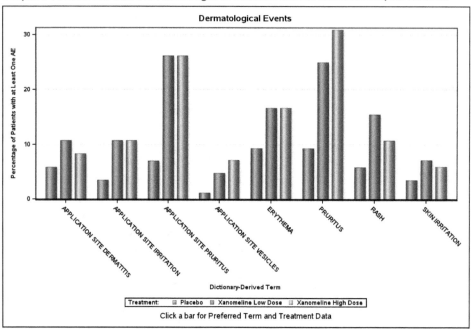

7.2.1.3 Creating a Drop-Down Selection Menu

If you have a SAS Stored Process and knowledge about SASjs, you can create a drop-down selection menu using the following example code available on GitHub - https://github.com/macropeople/minimal-seed-app. With the SASjs framework you can build great HTML web applications on SAS9 and SAS Viya. However, SASjs is beyond the scope of this book. You can also create a drop-down selection menu by using SAS, and a tiny proportion of HTML, JavaScript, and CSS code. You can see an example of a drop-down selection menu created primarily with SAS, and secondarily with HTML, JavaScript, and CSS here at on GitHub - https://github.com/rwatson724/SAS-Graphs-Clinical-Trials-Example. You will notice that as you choose a different severity level, the plot changes and displays the percentage of AEs that had the selected severity level. Using HTML, JavaScript, and CSS achieves this plot change. Output 7-8 shows you the different plots that can be selected from the drop-down menu.

In order to use the selection menu on your graphs, first you need to output all possible graphs, similar to the method used in Program 7-1. Therefore, for the AE severity graphs with the selection menu, four graphs are created. The four graphs are based on mild, moderate, and severe AEs, as well as all severities of AEs combined.

To be able to output these graphs, first, the counts for each severity are calculated, and then the percentages of patients that had at least one AE within that severity group are calculated. You can calculate the counts and percentages using PROC FREQ as shown in Program 7-3.

Display 7-3 shows an example of the data set used in order to create Output 7-8. Display 7-3 is very similar to Display 7-2, except Display 7-3 does not have the URL0 column, because there is no need for the individual bar charts to link to another page, and Display 7-3 has the additional columns AESEV, and AEDECOD2, which are discussed in more detail below:

- AESEV: the severities of the AEs, that is: mild, moderate or severe;
- AEDECOD2: the AE name with "#" embedded in some of the names. The use of "#" signals where the line will break. The "#" is for the tick values on the X axis to break at a chosen point, making the tick values take up less space on the Y axis compared to if the breaks were not there. Otherwise, the tick values will be very long, which will cause the values to take up more space in the layout area; thus, making the wall and data areas smaller. If you compare Output 7-7 with Output 7-8, you will notice that in Output 7-7 the tick values reduce the size of the wall and data area, which reduces the height of the Y axis. This is because the tick values in Output 7-7 display the tick values without any line breaks, and due to the tick values being quite long, they are rotated diagonally, which reduces the height of the area allotted, which affects the height of the Y axis.

Display 7-3: Data Snippet of Percentages of Patients with at Least One Adverse Event for Each Severity (SEVCNT2)

Obs	TRTA	AEDECOD	AESEV	COUNT	TOTCNT	PERCENT	AEDECOD2
2	Placebo	APPLICATION SITE DERMATITIS	MILD	5	86	5.8140	APPLICATION#SITE DERMATITIS
4	Placebo	APPLICATION SITE IRRITATION	MILD	2	86	2.3256	APPLICATION#SITE IRRITATION
7	Placebo	APPLICATION SITE PRURITUS	MILD	6	86	6.9767	APPLICATION#SITE PRURITUS
10	Placebo	APPLICATION SITE VESICLES	MILD	1	86	1.1628	APPLICATION#SITE VESICLES
12	Placebo	ERYTHEMA	MILD	6	86	6.9767	ERYTHEMA
15	Placebo	PRURITUS	MILD	8	86	9.3023	PRURITUS
18	Placebo	RASH	MILD	4	86	4.6512	RASH
21	Placebo	SKIN IRRITATION	MILD	2	86	2.3256	SKIN IRRITATION
24	Xanomeline High Dose	APPLICATION SITE DERMATITIS	MILD	6	84	7.1429	APPLICATION#SITE DERMATITIS
27	Xanomeline High Dose	APPLICATION SITE IRRITATION	MILD	8	84	9.5238	APPLICATION#SITE IRRITATION
30	Xanomeline High Dose	APPLICATION SITE PRURITUS	MILD	17	84	20.2381	APPLICATION#SITE PRURITUS
33	Xanomeline High Dose	APPLICATION SITE VESICLES	MILD	2	84	2.3810	APPLICATION#SITE VESICLES
36	Xanomeline High Dose	ERYTHEMA	MILD	12	84	14.2857	ERYTHEMA

Program 7-3 shows the HTML code that you need for the drop-down selection menu to work. It also shows you examples of how to output the graphs based on different severities of AEs. Program 7-3 uses the ODS HMTL5 statement instead of ODS HTML, which produces HTML 4. With the use of HTML 5, you can use some of the added benefits not available in HTML 4. One such benefit is the ability to embed the graph directly within the HTML page. With HTML 4, this feature is not available. Instead, a separate image file of the graph had to be created and then referenced in the HTML file in order for the graph to be displayed. Because of needing a separate image file for HTML 4, you needed to maintain more files.

Program 7-3: Creating Bar Charts with Different Severities

```
%let filterlink = <center><div class='dropdown'>
   <button onclick='myFunction()' class='dropbtn'>Select Severity
   </button>
   <div id='myDropdown' class='dropdown-content'>
      <a href='aes_ALL SEVERITIES.html'>All</a>  ❶
      <a href='aes_MILD.html'>Mild</a>  ❶
      <a href='aes_MODERATE.html'>Moderate</a>  ❶
       <a href='aes_SEVERE.html'>Severe</a>  ❶
   </div>
</div>
</center>;

%macro severity(severity, dynamiclabel);  ❷

proc template;  ❸
   define statgraph ptsevfreq;
      dynamic severity;  ❹
      begingraph;
         entrytitle textattrs = (size = 13pt) severity
             ": Dermatological Events";  ❹
         layout overlay /
            yaxisopts = (
                  label = "Percentage of Patients with at Least One AE"
                  labelattrs = (size = 11pt)
                  tickvalueattrs = (size = 9pt)
                  griddisplay = on
                  gridattrs = (color = lightgray pattern = dot)
                  linearopts = (tickvaluesequence = (start = 0 end = 30
                              increment = 5) viewmax = 33 viewmin = 0))

            xaxisopts = (labelattrs = (size = 11pt)
                  tickvalueattrs = (size = 9pt)
                  discreteopts = (tickvaluefitpolicy = splitalways  ❺
                  tickvaluesplitchar = "#"))  ❻;

            barchart category = AEDECOD2 response = PERCENT /
                              name = "dermevent"
                              group = TRTA groupdisplay = cluster
                              barwidth = 0.75 dataskin = sheen;

            discretelegend "dermevent" / title = "Treatment:"
                                    valueattrs=(size=10pt);

         endlayout;
      endgraph;
```

```
      end;
run;

ods html5 path = outp file = "aes_&severity..html"  ❼
      text="&filterlink."  ❽
      style = customsapphire headtext='<script src="myScript.js"> ❾
                                    </script>
         <link rel="stylesheet" type="text/css" href="myScript.css">';

proc sgrender data = tfldata.sevcnt2 template = ptsevfreq;
   where aesev = "&severity.";
   /* Pass severity level to the template. */
   dynamic severity = "&dynamiclabel.";  ❿
run;

ods html5 close;

%mend;
%severity(ALL SEVERITIES, %str());  ❷
%severity(MILD, %str(MILD :));  ❷
%severity(MODERATE, %str(MODERATE :));  ❷
%severity(SEVERE, %str(SEVERE :));
```

❶ The *FILTERLINK* macro variable stores the HTML code needed for the drop-down selection menu. In the HTML code, the **<a>** tag is used to link to the four outputs.

- **<a>** opens the anchor tag and **</a>** closes the anchor tag.

- The HREF attribute specifies the page to link to.

- The text in between the opening and closing tag is known as the link text or anchor text. This text is what you can see on the browser, for example, "All", "Mild", "Moderate", and "Severe".

- If you wanted to link to other outputs, you only need to edit the HREF attribute value, that is, the hyperlink and the link text within the anchor tag. The values would be of the format: <a href = 'ATTRIBUTE VALUE.HTML'>LINK TEXT </a>.

- If you need to change the drop-down menu heading within the <button tag>, this can simply be changed by replacing the button text "Select Severity" with another appropriate heading, which represents your drop-down menu heading within the <button> tag to another heading that represents your drop-down menu.

❷ The **SEVERITY** macro is created so that the graphs of all the severity levels can be created without having to write separate SGRENDER statements for each severity level. The macro has two parameters *SEVERITY* and *DYNAMICLABEL*. The *SEVERITY* parameter specifies the severity level to output, and the *DYNAMICLABEL* helps to specify the graph title.

❸ The PROC TEMPLATE code is very similar to the code in Program 7-2. Apart from using a DYNAMIC variable in the title and using tick value breaks on the X axis.

❹ The DYNAMIC statement creates the *SEVERITY* dynamic variable and the ENTRYTITLE statement displays the value of the *SEVERITY* dynamic variable in the title.

❺ The TICKVALUEFITPOLICY = SPLITALWAYS option always splits the axis tick value at every occurrence of a split character that is specified in the TICKVALUESPLITCHAR option.

❻ In this example, the TICKVALUESPLITCHAR = "#". Therefore, the tick value will start on a new line after it encounters a "#" symbol. Hence, the reasons why the "#" is incorporated into the AEDECOD2 values as shown in Display 7-3.

❼ ODS HTML5 FILE creates the AE graph of the mild severity in the HTML file aes_MILD.html.

❽ The ODS HTML5 statement contains a TEXT option, and values that are specified in the TEXT option are included in the body of the HTML file. TEXT= "&FILTERLINK", adds the severity drop-down selection menu to the file, and as a result, to the graph. Without the TEXT = "&FILTERLINK" on each file, you would not be able to navigate to the other severity graphs.

❾ The HEADTEXT option in the ODS HTML5 statement writes text to the head of the HTML file. You can use the HEADTEXT option to reference the external CSS file and JavaScript files that you want to use in your program. An HTML file contains a **<head>** tag and a **<body>** tag. The head of an HTML document is the part that is not displayed in the web browser when the page is loaded. The head contains information such as the page <title>, links to CSS files, links to JavaScript files, and other metadata, such as the author, and relevant keywords that describe the document (MDN contributors, 2020).

❿ The DYNAMIC SEVERITY = "*&DYNAMICLABEL*" statement within PROC SGRENDER, assigns the *DYNAMICLABEL* macro variable value to the *SEVERITY* dynamic variable. For example, for the MILD AEs, the value "MILD :" is assigned. With the use of the DYNAMIC statement in PROC TEMPLATE, this allows the dynamic variable *SEVERITY* to be used in other GTL statements, as demonstrated with the use of the ENTRYTITLE statement as seen in callout **❸**.

You can see the graphs of the AE bar charts for each severity in Output 7-8. Output 7-8 shows a large graph of the AE bar chart with the severity drop-down menu box. When you click on one of the severity options, such as "Mild", you are then shown the graph of the patients that had mild adverse events. The four smaller AE bar charts illustrate the plots that you see after clicking on each of the severity selections. The first plot contains all the AE severity levels, and the second, third and fourth plot, contains the patients that had the mild, moderate, and severe AEs respectively.

Output 7-8: Severity Drop-Down Selection Menu Graphs

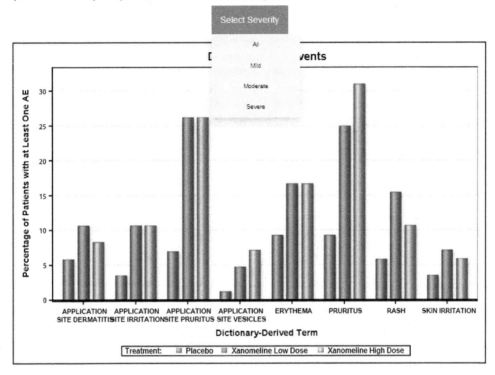

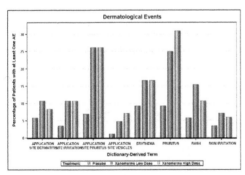

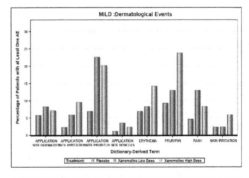

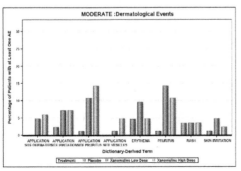

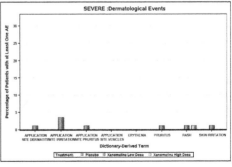

7.3 Venn Diagrams

John Venn (1834–1923), the Hull-born philosopher and mathematician, introduced Venn diagrams in 1883. Venn Diagrams are a great way to visualize elements that are unique to only one group and simultaneously visualize elements intersecting with other groups. Venn diagrams are symmetrical by nature, although when the number of groups is quite high, such as more than 5, some loss of symmetry in the diagrams is unavoidable. A Venn diagram in which the area of each shape is proportional to the number of elements that it contains is known as a proportional Venn diagram. Proportional Venn diagrams are generally not symmetrical. This section focuses on symmetrical Venn diagrams. The number of groups in a Venn diagram = 2^n (including the group outside the diagram).

It is possible to produce Venn diagrams using the software JMP® Statistical Discovery™ from SAS. You can also use the web applications (Oliveros, 2018 and Hulsen, 2018) to create Venn diagrams. However, you might need or want to keep your data within SAS, and this section shows you how you can use SAS to create 2-, 3-, and 4- way Venn diagrams.

7.3.1 Creating a Two-Way Venn Diagram

If you wanted to create Venn Diagrams to assess the AEs that are common and different between the xanomeline low dose and high dose groups, first you need to identify the distinct values of AEs that are in the low dose group, the high dose group, and then merge the two columns together. You can do this using PROC SQL or the MERGE statement within a DATA step. Regardless of which approach you use to join the data; the results would look similar to Display 7-4.

Display 7-4: Data Snippet of Adverse Events for Xanomeline Low Dose and Xanomeline High Dose*

Obs	placebo_AEs	xan_low_AEs	xan_high_AEs
1	ABDOMINAL PAIN	ABDOMINAL PAIN	ABDOMINAL DISCOMFORT
2	AGITATION	AGITATION	ABDOMINAL PAIN
3	ALOPECIA	ANXIETY	ACROCHORDON EXCISION
4	ANXIETY	APPLICATION SITE BLEEDING	ACTINIC KERATOSIS
5	APPLICATION SITE DERMATITIS	APPLICATION SITE DERMATITIS	AGITATION
6	APPLICATION SITE ERYTHEMA	APPLICATION SITE DESQUAMATION	ALCOHOL USE
7	APPLICATION SITE INDURATION	APPLICATION SITE DISCOLOURATION	ALLERGIC GRANULOMATOUS ANGIITIS
8	APPLICATION SITE IRRITATION	APPLICATION SITE ERYTHEMA	AMNESIA
9	APPLICATION SITE PRURITUS	APPLICATION SITE IRRITATION	APPLICATION SITE DERMATITIS
10	APPLICATION SITE REACTION	APPLICATION SITE PRURITUS	APPLICATION SITE DISCHARGE

*This sample data set is not provided. It is just shown for illustration purposes.

Once you have your input data set ready, you can use the macro, %***VennDiagramMacro***, which is available at https://github.com/rwatson724/SAS-Graphs-Clinical-Trials-Example, to calculate the numbers of AEs that are in both the low dose and high dose groups, and the AEs that are just in either the low dose group or the high dose group. You can also use the macro to produce the

Venn diagram for you. This section aims to demonstrate the GTL code that you can use to produce the Venn diagrams, and Program 7-4, shows you how to create a two-way Venn diagram. The DRAWOVAL statement is significant for creating Venn diagrams.

Program 7-4: Creating Bar Charts with Different Severities

```
data data_for_plot_layout;    ❶
   do x = 1 to 100;
      y = x;
      output;
   end;
run;

proc template;
   define statgraph Venn2Way;
      begingraph / drawspace = datavalue;    ❷
         /* Plot */
         layout overlay / yaxisopts = (display = NONE)    ❸
                          xaxisopts = (display = NONE);    ❸

            scatterplot x = x y = y / markerattrs = (size = 0);    ❹

             /* Venn Diagram (Circles) */
             drawoval x = 37 y = 50 width = 45 height = 60 /    ❺
                display = all    ❻
                fillattrs = (color = red)    ❼
                transparency = 0.75    ❽
                widthunit = percent heightunit = percent;    ❾

             drawoval x = 63 y = 50 width = 45 height = 60 /
                display = all
                fillattrs = (color = green)
                transparency = 0.75
                widthunit = percent
                heightunit = percent;

              /* Numbers */
              drawtext "61" / x = 33 y = 50 anchor = center;    ❿
              drawtext "63" / x = 50 y = 50 anchor = center;    ❿
              drawtext "66" / x = 66 y = 50 anchor = center;    ❿

              /* Labels */
              drawtext "Xanomeline Low Dose" /
                 x = 30 y = 15 anchor = center width = 32;    ❿

              drawtext "Xanomeline High Dose" /
                 x = 70 y = 15 anchor = center width = 32;    ❿

         endlayout;
      endgraph;
   end;
run;

proc sgrender data=data_for_plot_layout template=Venn2Way;
run;
```

❶ A data set is created with 2 variables and 100 observations. The two variables are X and Y, and they go from the numbers 1 to 100. The data set is used to produce the data area for the plot. The reason the data set is used is because you need to associate a data set to with the template when you use PROC SGRENDER. The numbers in the data set range from 1 to 100, because the Venn diagram and the associated labels have coordinates between the values of 1 and 100. If the data set only had two observations, with the first observation being 1, and the second observation being 100, this would work correctly too, because the Venn diagram and the labels are within the minimum and maximum values within the data, and hence the axes. If the data set only had one observation with X = 1 and Y = 1 and DRAWSPACE = DATAVALUE option is used, then in order to see the Venn diagram, you would need to include the VIEWMAX option and specify a large value such as 100, within the XAXISOPTS and YAXISOPTS options. Although, if any of the draw space options that represent the percentages are used, such as GRAPHPERCENT, LAYOUTPERCENT, WALLPERCENT, and DATAPERCENT, then only having a data with one observation would be fine.

❷ The DRAWSPACE option in the BEGINGRAPH statement specifies the drawing space and drawing units for the draw statements. DRAWSPACE = DATAVALUE specifies that the drawing space is within the actual graph, and that the points should be drawn at the exact values. That is, if there is a value of X = 30 in the DRAWTEXT statement, then the text will be located where X = 30 on the X axis of the graph.

❸ On the Venn diagram outputs, there is no need to have any axis labels, lines, tick values, and tick marks, and therefore, display = NONE is used in both the XAXISOPTS and YAXISOPTS.

❹ The SCATTERPLOT statement produces a plot with markers on a diagonal line going from 1 to 100. The main aim of using the SCATTERPLOT statement is to create a blank canvas for the Venn diagrams to be overlaid. Therefore, SIZE = 0 marker attributes option is used to hide the markers and display the blank canvas. You can see how the SCATTERPLOT would have looked and how the tick values would have been if they were not hidden in Output 7-12, which is in the Appendix to this chapter. The only differences with the code in Program 7-4 and the code to produce Output 7-12, which is shown in the Appendix to this chapter, are:

● In the XAXISOPTS and YAXISOPTS options, the DISPLAY = (TICKVALUES TICKS) options were used to display the tick values and ticks on the graphs from 0 to 100.

● In the SCATTERPLOT statement, the MARKERATTRS = (SIZE = 1) was used to make the markers visible.

● In the DRAWTEXT the width was increased using WIDTH = 36, so that the treatment labels were on one line. This was to compensate for the extra space taking up from Adding the X- and Y-axis tick values.

If you use DRAWSPACE = DATAVALUE within the BEGINGRAPH statement, then a SCATTERPLOT or other plot statement is required in order to be able to use the DRAWOVAL and DRAWTEXT statements. If the GRAPHPERCENT or LAYOUTPERCENT drawspace was used, then the SCATTERPLOT statement could be omitted and the Venn diagrams will still be displayed fine.

❺ The DRAWOVAL statement indicates the location and the size of the Venn diagrams. X = 37 and Y = 50 specifies that the center of the Venn diagrams will start at those coordinates. WIDTH = 45 and HEIGHT = 60 indicate the size of the Venn diagram. The specified locations and sizes for the Venn diagrams were found using trial and error. There is probably a more mathematical way to draw the Venn diagrams; however, the specified options are sufficient. The reason the height is larger than the width by one-third is because when the Venn diagrams are being outputted using the default width and height. With the default width (640px = 6.67 inches) and height (480px = 5 inches), the width of the graph is one-third bigger than the height. Therefore, making the height one-third larger here in the DRAWOVAL statement ensures that circles instead of ovals are drawn as Venn diagrams. If you want to produce a two-way Venn diagram and you are using the default width and height output options, or you are using output options where the width is one-third larger than the height, then using the DRAWOVAL options that are specified here is recommended. This is because the circles are a sufficient size and are positioned in a good location.

You can also use the same value of WIDTH and HEIGHT in the DRAWOVAL statements, that is, WIDTH = 45 and HEIGHT = 45, and still achieve circles for the Venn diagrams even though the output width is one-third larger than the output height, by using the OVERLAYEQUATED layout, with the EQUATETYPE = SQUARE option. With the OVERLAYEQUATED layout, the X and Y axes display units are the same. You can see a plot of the two-way Venn diagram using the same WIDTH and HEIGHT in the DRAWOVAL statement and the OVERLAYEQUATED layout in Output 7-13 within the Appendix to this chapter.

❻ The DISPLAY = ALL options specifies to display both an outlined oval and a filled-in oval.

❼ COLOR = RED fill attributes option fills the Venn diagram in red.

❽ The TRANSPARENCY option specifies the amount of transparency, where 0 is opaque, and 1 is completely transparent.

❾ The WIDTHUNIT = PERCENT and HEIGHTUNIT = PERCENT specifies that the value in WIDTH and HEIGHT should be considered as a percentage of the data area, as opposed to the number of pixels, or the data values.

❿ The DRAWTEXT statement displays the number of unique AEs in each group and the group names in the Venn diagrams. The X option specifies the X coordinate of text, and the Y option specifies the Y coordinate of text. The ANCHOR option indicates where the text will be drawn with respect to the coordinates. The WIDTH option specifies the width of the text. If a small value is used for the WIDTH option, then the text might continue on a separate line. Note that the values in the first three DRAWTEXT statement can be determined programmatically and use a macro variable to populate rather than enter the specific value on the statement. The specific counts are entered in the DRAWTEXT statements for illustration purposes so that you can see where the values are being "drawn" on the graph in Output 7-9.

Output 7-9 shows a two-way Venn diagram of the relationship between the investigative product treatments taken and the AEs. You can see that 63 AEs occurred in both treatment groups. Sixty-one AEs were only present in patients treated with the low dose, and 66 AEs were only present in patients treated with the high dose.

Output 7-9: Two-Way Venn Diagram of the Relationship Between Treatment and Adverse Events

7.3.2 Creating a Three-Way Venn Diagram

To produce three-way Venn diagrams, all that is needed is to add the third variable and label to the existing two-way Venn diagram GTL code. Essentially, you only need to add a new DRAWOVAL statement for the third circle, and four more DRAWTEXT statements for the additional numbers. These new statements account for the new treatment group and the interactions between the third treatment with the other two treatment groups. Finally, an extra DRAWTEXT statement is needed to create the label for the new treatment group.

Program 7-5: Three-Way Venn Diagram

```
data data_for_plot_layout;
   do x = 1 to 100;
      y = x;
      output;
   end;
run;

proc template;
   define statgraph Venn3Way;
      begingraph / drawspace = datavalue;
      /* Plot */
      layout overlay / yaxisopts = (display = none) xaxisopts = (display = none);
         scatterplot x = x y = y / markerattrs = (size = 0);

            /* Venn Diagram (Circles) */
            drawoval x = 37 y = 40 width = 45 height = 60 /  ❶
               display = all fillattrs = (color = red)
```

```
            transparency = 0.75
            widthunit = percent heightunit = percent;

        drawoval x = 63 y = 40 width = 45 height = 60 /    ❶
            display = all fillattrs = (color = green)
            transparency = 0.75
            widthunit = percent heightunit = percent;

        drawoval x = 50 y = 65 width = 45 height = 60 /    ❶
            display = all fillattrs = (color = blue)
            transparency=0.75 widthunit= percent heightunit= percent;

        /* Numbers */
        drawtext "48" / x = 32 y = 35 anchor = center;    ❷
        drawtext "22" / x = 50 y = 30 anchor = center;    ❷
        drawtext "52" / x = 68 y = 35 anchor = center;    ❷
        drawtext "41" / x = 50 y = 50 anchor = center;    ❷
        drawtext "13" / x = 37 y = 55 anchor = center;    ❷
        drawtext "14" / x = 63 y = 55 anchor = center;    ❷
        drawtext "52" / x = 50 y = 75 anchor = center;    ❷

            /* Labels */
        drawtext "Xanomeline Low Dose" / x = 30 y = 7
                                    anchor = center width = 33;
        drawtext "Xanomeline High Dose" / x = 70 y = 7
                                    anchor = center width = 33;
        drawtext "Placebo" / x = 50 y = 98 anchor = center width = 33;
      endlayout;
    endgraph;
  end;
run;

proc sgrender data=data_for_plot_layout template=Venn3Way;
run;
```

❶ Compared to Program 7-4, there is one more DRAWOVAL statement, which draws the third circle. The first and second circles in the Venn diagrams, that is, the Low Dose and High Dose circles have the same values for the X coordinates as in Program 7-4. However, the Y coordinates are now lower, and are now at Y = 40 instead of Y = 50. The lower Y coordinates helps the third circle to fit into the Venn diagram, because the Y axis of the Venn diagram spans from 0 to 100. The third circle is drawn in the middle of the X axis at X = 50. The Y coordinate of the third circle is at Y = 65.

❷ Compared to Program 7-4, there are three more DRAWTEXT statements, which displays the numbers in the additional groups.

Output 7-10 shows a three-way Venn diagram. You can see that the number of AEs are very similar for each treatment group, and you can see that 41 AEs occurred in all three treatment groups.

Sixty-three AEs occurred in both low dose and high dose treatment groups, which is calculated by summing the values 41 (AEs in all three groups) and 22 (AEs in low dose and high dose only). Fifty-four AEs occurred in both low dose and placebo treatment groups. Fifty-five AEs occurred in both high dose and placebo treatment groups.

Output 7-10: Three-Way Venn Diagram of the Relationship Between Treatment and Adverse Events

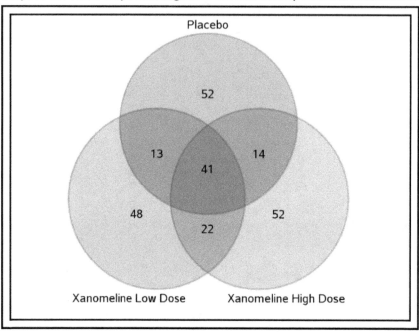

7.3.3 Creating a Four-Way Venn Diagram

Now you have seen how to produce two-way and three-way Venn diagrams, you will see how to produce four-way Venn diagrams. The four-way Venn diagram needs an additional DRAWOVAL statement and more DRAWTEXT statements. The four-way Venn diagram displays ellipses instead of circles, and to produce the ellipses, you can use a ratio of 1:4 between the WIDTH and the HEIGHT in the DRAWOVAL statements. You can use the ROTATE argument to position the ellipses in the right direction and the right degree of rotation so that the ellipses are not on top of each other. Program 7-6 demonstrates how to create a four-way Venn diagram.

Program 7-6: Creating Bar Charts with Different Severities

```
data data_for_plot_layout;
   do x = 1 to 100;
      y = x;
      output;
   end;
run;

proc template;
   define statgraph Venn4Way;
      begingraph / drawspace = datavalue;
      /* Plot */
      layout overlay / yaxisopts = (display = none)
                       xaxisopts = (display = none);
         scatterplot x = x y = y / markerattrs = (size = 0);
```

```
            /* Venn Diagram (Circles) */
        drawoval x = 28 y = 39 width = 26 height = 100 /  ❶
           display = all
           fillattrs = (color = red transparency = 0.85)  ❷
           outlineattrs = (color = red)  ❸
           transparency = 0.50  ❹
           widthunit = percent
           heightunit = percent
           rotate = 45;  ❺

        << SAS CODE for additional DRAWOVAL statements>>

         /* Numbers */
        drawtext textattrs = graphvaluetext(size = 6pt weight = bold)
             "20" /  ❻   x = 13 y = 60 anchor = center;

        << SAS CODE for additional DRAWTEXT statements>>

        /* Labels */
        drawtext textattrs = graphvaluetext(size = 7pt weight = bold)
        "Low Dose - Males" / x = 13 y = 20 anchor = center width = 30;

        << SAS CODE for additional DRAWTEXT statements>>

      endlayout;
    endgraph;
  end;
run;

ods graphics / reset=all height=3.75in width=5in;

proc sgrender data=data_for_plot_layout template = Venn4Way;
run;
```

❶ The width is almost four times smaller than the height to help draw the ellipse shape.

❷ To distinguish between the different ellipses, you can use the FILLATTRS option to specify the color of the ellipse to be drawn. In addition, you can use TRANSPARENCY fill attributes option to specify the amount of transparency for the fill color, where 0 is opaque, and 1 is completely transparent. The effect of this option also depends on the TRANSPARENCY option in callout ❹, which controls the transparency options of both the outline and the fill color. In order to see the effect of the transparency fill options, a value of 1 should not be chosen for the TRANSPARENCY option in callout ❹, because this would make the whole ellipse transparent, no matter the value in the TRANSPARENCY fill attributes option. To create an ellipse that had a solid red outline but did not have any fill color, you would use the options FILLATTRS = (COLOR = RED TRANSPARENCY = 1) TRANSPARENCY = 0.

The first ellipse in Program 7-6 has the options FILLATTRS = (COLOR = RED TRANSPARENCY = 0.85) TRANSPARENCY = 0.5. First, a transparency of 0.5 is set on the outline and fill color, followed by a transparency of 0.85 on the original transparency of 0.5. You could have achieved the same fill attribute transparency effect by using the options FILLATTRS = (COLOR = RED TRANSPARENCY = 0.925) TRANSPARENCY = 1. You can also think of the final actual fill color transparency of being:

- TRANSPARENCY + (1 − TRANSPARENCY) * TRANSPARENCY FILL ATTRIBUTE.

- First DRAWOVAL transparency is (0.5 + (1 − 0.5) * 0.85 = 0.925.

- FILLATTRS = (COLOR = RED TRANSPARENCY = 0.85) TRANSPARENCY = 0.5, can also be specified as FILLATTRS = (COLOR = RED TRANSPARENCY = **0.925**) TRANSPARENCY = 1.

❸ The OUTLINEATTRS = (COLOR = RED) option specifies that the outline of the first ellipse drawn is red. With the use of OUTLINEATTRS, you can specify different colors for the outline. Such as, OUTLINEATTRS = (COLOR = YELLOW), OUTLINEATTRS = (COLOR = BLUE), and OUTLINEATTRS = (COLOR = GREEN) specify that the outlines of the second, third, and fourth ellipses are drawn in yellow, blue, and green, respectively.

❹ The TRANSPARENCY option specifies the amount of transparency in the outline and the fill color, where 0 is opaque, and 1 is completely transparent. You can set a different transparency for the fill color compared to the outline color by using the TRANSPARENCY fill attributes option in callout ❷.

❺ The ROTATE option specifies the angle of the ellipses. In other words, it specifies the degree of rotation.

❻ The TEXTATTRS = GRAPHVALUETEXT(SIZE = 6PT WEIGHT = BOLD) option specifies that the text labels use the default format, such as the default font. However, they should be bold and be in size 6pt. The names have a size of 6pt so that they can be smaller, and this is because the four-way Venn diagram has a lot of information to display. Bold is specified to ensure that even though the text is smaller, it is still readable. GRAPHVALUETEXT inherits the default style attributes.

Output 7-11 shows a four-way Venn diagram of the relationship between the treatments taken, the gender, and the AEs. You can see that 16 AEs occurred in the low dose and high dose treatments, and in both males and females.

Output 7-11: Four-Way Venn Diagram of the Relationship Between Treatment, Gender, and Adverse Events

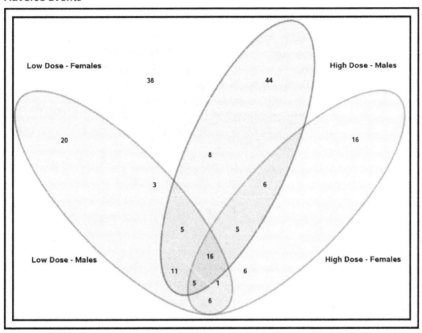

7.4 Chapter Summary

In this chapter, you saw how to produce interactive graphs and Venn diagrams. Creating interactive graphics can make your figures more centralized because you only need to access them from one location. With the use of interactive graphs, it is easier to identify patterns by having the ability to drill down on a specific component versus looking at separate graphs and trying to align them.

Producing interactive graphs is quite similar to producing standard graphs. The only difference is that you are creating HTML outputs, using the IMAGEMAP = ON option in the ODS GRAPHICS statement, and using the URL option in the plot statements.

As well as clicking on a bar to create drill-down interactivity, you can also create a drop-down selection menu. To create the drop-down selection menu, also requires some HTML code, JavaScript, and CSS. However, the JavaScript and CSS code are readily available online, and if you want to create drop-down selection menus, then you only would need to learn a little HTML.

Venn diagrams are handy for visualizing the relationships between groups, such as the number of AEs which each treatment group experienced. Venn diagrams might not be easy to produce, but they can provide insight into how items are related and unrelated by showing the overlap between groups, and the items exclusive to one group. This chapter has demonstrated how you can implement Venn Diagrams using SAS 9.4, with aesthetically pleasing results.

References

Harris, K., and Watson, R. (2019). "Interactive Graphs." Proceedings of the Annual Pharmaceutical SAS Users Group Conference. Cary, NC: SAS Institute Inc. *PharmaSUG*. Philadelphia: PharmaSUG.

Hulsen, T. (2018, 3 1). *BioVenn*. Retrieved from http//www.biovenn.nl/index.php

MDN web docs. (2020, July 05). *What's in the head? Metadata in HTML*. Retrieved from Resources for developers, by developers: https://developer.mozilla.org/en-US/docs/Learn/HTML/Introduction_to_HTML/The_head_metadata_in_HTML

Oliveros, J. C. (2018, 03 01). *Venny 2.1*. Retrieved from http://bioinfogp.cnb.csic.es/tools/venny/index.html

Appendix

JavaScript

```
function myFunction() {
  document.getElementById("myDropdown").classList.toggle("show");
}

// Close the dropdown menu if the user clicks outside of it
window.onclick = function(event) {
  if (!event.target.matches('.dropbtn')) {
    var dropdowns = document.getElementsByClassName("dropdown-content");
    var i;
    for (i = 0; i < dropdowns.length; i++) {
      var openDropdown = dropdowns[i];
      if (openDropdown.classList.contains('show')) {
        openDropdown.classList.remove('show');
      }
    }
  }
}
```

CSS

```
/* Dropdown Button */
.dropbtn {
  background-color: #3498DB;
  color: white;
  padding: 16px;
  font-size: 16px;
  border: none;
  cursor: pointer;
}

/* Dropdown button on hover & focus */
.dropbtn:hover, .dropbtn:focus {
  background-color: #2980B9;
}

/* The container <div> - needed to position the dropdown content */
.dropdown {
  position: relative;
  display: inline-block;
}

/* Dropdown Content (Hidden by Default) */
.dropdown-content {
  display: none;
  position: absolute;
  background-color: #f1f1f1;
  min-width: 160px;
  box-shadow: 0px 8px 16px 0px rgba(0,0,0,0.2);
  z-index: 1;
}
```

```
/* Links inside the dropdown */
.dropdown-content a {
  color: black;
  padding: 12px 16px;
  text-decoration: none;
  display: block;
}

/* Change color of dropdown links on hover */
.dropdown-content a:hover {background-color: #ddd}

/* Show the dropdown menu (use JS to add this class to the .dropdown-
content container when the user clicks on the dropdown button) */
.show {display:block;}
```

Output 7-12: Two-Way Venn Diagram of the Relationship Between Treatment and Adverse Events with Tick Values

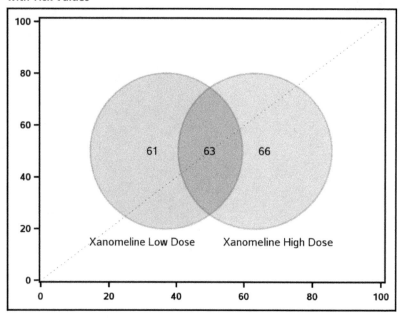

Output 7-13: Two-Way Venn Diagram of the Relationship Between Treatment and Adverse Events Using LAYOUTOVERLAYEQUATED

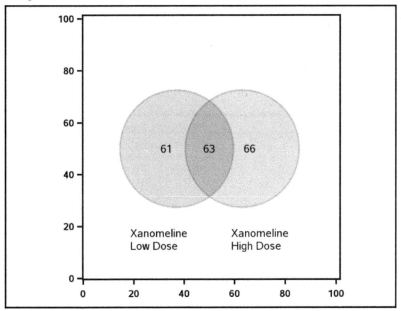

Chapter 8: Building the Seemingly Impossible

8.1 Introduction to Patient Profile Plot

Most of the graphs that you produce are pretty straight forward with maybe one or two types of data. But what if you are asked to produce a patient profile plot? If you have never produced a patient profile plot before, you might ask yourself a series of questions.

- What is a patient profile plot?

- What type of data should be captured?

- What is the best approach for producing one?

A patient profile plot is a visual representation of the patient's data. It is used to help obtain a better understanding of that particular patient's data and can sometimes be used to help determine whether a patient has their medical condition under control or whether they are using medication or some intervention to keep it under control.

A patient profile plot typically contains several data sources that showcase the patient's journey through the study. Typically, the data captured on the profile plot is the data that is key to understanding the condition that is considered the primary or secondary indication for the study.

The best approach for producing a profile plot depends on the type of data captured as well as what you are trying to convey. A patient profile plot can be a number of mini plots displayed on the same page or it can be one plot with a variety of data. In this chapter, a step-by-step approach is shown for a diabetic profile plot in one plot.

8.1.1 Walking Through a Diabetic Patient Profile Plot

What if you were asked to produce a plot like the one shown in Output 8-1? Would you even know where to begin? When encountered with something that seems almost impossible to do at first glance, take a step back and examine each component individually and start with what you know.

Output 8-1: Diabetic Patient Profile Plot for Patient 0001

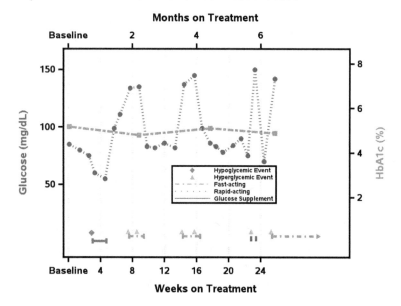

8.1.1.1 Creation of Data Set with All Necessary Data to Build a Diabetic Patient Profile Plot

Regardless if you have a template that was created by another or you are creating your own template, you will need to ensure that the data is in a structure that the template can use. For the walk-through of the creation of the diabetic profile plot, you need to create a new template. The data might not be in a structure that is readily available to create the desired graph. Therefore, you need to think about the best approach for creating the data set. Program 8-1 illustrates the creation of the data set in a structure that can be used by the template. Display 8-1 provides a portion of the data that is produced by Program 8-1.

Program 8-1: Program to Structure the Data in a Format to be Used by Custom Template

```
options validvarname = upcase;
libname adam "<location of permanent SAS data sets>";

proc sort data = adam.adlbdiab  out = adlbdiab;
   by usubjid ady paramcd;
   where anl01fl = 'Y';  ❶
run;

proc transpose data = adlbdiab  ❷  out = tadlb (drop = _:);
   by usubjid ady adt;
   var aval;
   id paramcd;
run;

data all;  ❸
      set tadlb  adam.adaediab (drop = trtsdt)
          adam.adcmdiab (drop = cmtrt trtsdt);

   strtday = coalesce(ady, astdy);  ❹

   /* create barbed arrow for ongoing events */
   length cmcap $15;  ❺
   if cmenrtpt = 'ONGOING' then cmcap = 'FILLEDARROW';
   else cmcap = 'SERIF';
run;

/* sort reverse order so that can find last study day for ongoing events */
proc sort data = all;
   by usubjid descending strtday;
run;

/* determine the last study day for ongoing events */
data adam.diabprof;
   set all;
   by usubjid descending strtday;
   retain endday;
   if first.usubjid then endday = .;

   if ady ne . then endday = ady;  ❻
   else if (aedecod ne '' or acat ne '') and aendy ne . then
       endday = aendy;
   else if (aedecod ne '' and aeenrtpt = 'ONGOING') or
           (acat ne '' and cmenrtpt = 'ONGOING') then endday = endday + 20;

   if index(aedecod, 'Hypo') then event1 = 8;    ❼
   else if index(aedecod, 'Hyper') then event2 = 9;

   if index(acat, 'Supplement') then med = 1;    ❽
   else if index(acat, 'Rapid') then med = 3;
   else if index(acat, 'Fast') then med = 5;
   else if index(acat, 'Elev') then med = 7;
run;
```

❶ Because there can be multiple records for a subject, parameter, and time point combination, you need to make sure you are selecting only the lab records that are used for analysis (ADLBDIAB where ANL01FL = "Y").

❷ Since two lab results are of interest: Glucose and HbA1c, then these values need to be on the same record; therefore, the data is transposed.

❸ Based on the plot, you need to also include information about hyperglycemic and hypoglycemic events, which comes from the adverse event data set (ADAEDIAB) as well as information about the type of rescue medication the patient might have taken, which comes from the concomitant medication data set (ADCMDIAB).

❹ Due to the nature of the different data sources, ADLBDIAB captures the study day in ADY but ADAEDIAB and ADCMDIAB have a study start (ASTDY) and end (AENDY) day. Therefore, a variable needs to be created that will capture the start day. Note that since the data sets are being combined both ADY and ASTDY will not be populated at the same time. With the COALESCE function, the first nonmissing value in the list of arguments is used to populate the variable.

❺ CMCAP is a variable that will be used in the template to indicate whether a medication is considered ongoing. The values correspond to possible values that can be used for GTL plots that have the LOWCAP or HIGHCAP option.

❻ A new variable, ENDDAY, is created based on whether the end day already exists. For lab data, ADY is used to represent the end day. For AE and medication data, AENDY is used to represent the end day. If the end day does not exist and the event or the medication is considered ONGOING, then the end day of the previous record is used, and 20 days is added so that it extends beyond the last day, and an arrow can be used to indicate it is ongoing.

❼ Each event is captured as a separate variable with a numeric value. Hypoglycemic events are denoted as EVENT1 = 8; while hyperglycemic events are denoted as EVENT2 = 9. Although you could create one variable to distinguish between the two events and assign them different numeric values (for example, EVENT = 8 for hypoglycemic and EVENT = 9 for hyperglycemic), it would be more difficult to control the color and symbols used to represent each event. Therefore, in order to get them to appear with different markers and different colors, each event had its own numeric variable. The values of 8 and 9 were chosen because you want the events to appear at the bottom of the graph and based on the lab data, you know that glucose ideally should not get below 10 mg/dL.

❽ The medications used when a patient had an event are also assigned a numeric code. Unlike the events, one variable is needed to represent the different types of medication. The reason for this is explained in Section 8.1.1.4 *Creation of Highlow Plot for Duration of Medication by Type of Medication*.

Display 8-1: Portion of the Lab Data Produced by Program 8-1

Obs	USUBJID	ADY	ADT	HBA1C	SGLUC	AEDECOD	ASTDT	ASTDY	AENDT	AENDY
1	ABC-DEF-0001	180	03MAY2017	4.88	142		.	.	.	.
2	ABC-DEF-0001	.	.	.	.	Hyperglycemia	30APR2017	177	.	.
3	ABC-DEF-0001	.	.	.	.		30APR2017	177	.	.
4	ABC-DEF-0001	171	24APR2017	.	70		.	.	.	.
5	ABC-DEF-0001	163	16APR2017	.	150		.	.	.	.
6	ABC-DEF-0001	.	.	.	.	Hyperglycemia	13APR2017	160	15APR2017	162

Obs	USUBJID	ADY	ADT	HBA1C	SGLUC	AEDECOD	ASTDT	ASTDY	AENDT	AENDY
7	ABC-DEF-0001	.	.	.	.		12APR2017	159	18APR2017	165
8	ABC-DEF-0001	157	10APR2017	.	75		.	.	.	.
9	ABC-DEF-0001	151	04APR2017	.	90		.	.	.	.
10	ABC-DEF-0001	144	28MAR2017	.	84		.	.	.	.

Obs	AEENRTPT	CMDECOD	ACAT	CMENRTPT	STRTDAY	CMCAP	ENDDAY	EVENT1	EVENT2	MED
1					180	SERIF	180	.	.	.
2	ONGOING				177	SERIF	200	.	9	.
3		FlexPen	Fast-acting	ONGOING	177	FILLEDARROW	220	.	.	5
4					171	SERIF	171	.	.	.
5					163	SERIF	163	.	.	.
6					160	SERIF	162	.	9	.
7		NovoLog	Rapid-acting		159	SERIF	165	.	.	3
8					157	SERIF	157	.	.	.
9					151	SERIF	151	.	.	.
10					144	SERIF	144	.	.	.

Once you have the data in the desired format to align with the needs to produce the graph, you can create a template that uses the variables and structure of the data set or if a template already exists you can associate the data with the template to render the graph. As mentioned previously for the purpose of this walk-through, you assume that no template exists and you need to create your own.

8.1.1.2 Creation of Series Plots for Glucose and HbA1c Lab Results

Upon examining the plot that you need to create, you notice that there are two series plots. Series plots are pretty straight forward, so you are able to create those using the SERIESPLOT statement in GTL. Although it would be easy enough to create the series plot using SGPLOT, you want to build your own custom template because the final plot contains more than a series plot. Before you create the series plots you need to determine what the axes will be. Looking at the data you notice that Glucose (mg/dL) levels are captured approximately every week; however, HbA1c (%) is captured approximately every two months. Therefore, not only do the two lab tests have different lab value ranges, they also have different time points. Let us start with creating one plot at a time. Program 8-2 uses the data created in Program 8-1 to generate a series plot for Glucose, as seen in Output 8-2.

Program 8-2: Template to Produce Series Plot for Glucose (mg/dL)

```
proc format;
   picture dy2wk (round)     ❶
              0 = 'Baseline' (noedit)
       1 - high = 09 (mult = 0.142857);
run;

proc template;
   define statgraph glucprof;
      begingraph / border = false;
         layout overlay / xaxisopts = (label = 'Weeks on Treatment'   ❷
                          griddisplay = on     ❸
                          gridattrs = (pattern = 35 color = libgr)     ❹
                          labelattrs = (color = black weight = bold)
                             linearopts = (viewmin = 0    viewmax = 180
                                 tickvaluesequence = (start = 0 end = 168
                                                          increment = 28)
                                     tickvalueformat = dy2wk.))

                          yaxisopts = (label = 'Glucose (mg/dL)'    ❺
                             offsetmin = 0.07 offsetmax = 0.07
                             labelattrs = (color = firebrick
                                            weight = bold)
                             linearopts = (viewmin = 0 viewmax = 155));

            seriesplot x = strtday y = sgluc / display = all   ❻
               markerattrs = (symbol = circlefilled   ❼
                              color = firebrick size = 6)
               lineattrs = (color = firebrick pattern = 34   ❽
                             thickness = 3);
         endlayout;
      endgraph;
   end;
run;

proc sgrender data = adam.diabprof template = glucprof;
   where usubjid = 'ABC-DEF-0001';   ❾
run;
```

❶ In order to preserve the value of the study day, a picture format is created so that the time point displays as weeks instead of days.

❷ Within the layout specified, you can indicate which axes are needed and use the various options associated with the axis to customize. Because Glucose is captured weekly, the label for the X axis is set to indicate "Weeks" and options are set to indicate that only values between study day 0 and 180 are to be displayed but the X axis will only display tick marks from 0 to 168 in increments of 28 days. The format that was created is applied to the tick marks to convert it from days to weeks.

❸ Sometimes gridlines on the graph help to see where a particular point falls with respect to the values on the axis. In order to display, you can use GRIDDISPLAY = ON.

❹ Each ODS style has different graph style elements that control the appearance of different aspects of the graph. The GraphGridLines style element is used for the appearance of the gridlines. In order, to modify the appearance of the gridlines, you can use GRIDATTRS option. Within the GRIDATTRS option, you can control attributes such as the line pattern and line color.

❺ Similar to the X axis, the Y axis can be customized to provide a specific label as well as label attributes such as color. If OFFSETMIN and OFFSETMAX options are used on the Y axis, it signals how much space needs to be reserved from the bottom axis and top axis respectively. In addition, these options can be used on the X axis as well. If used on the X axis, then it reserves space from the left and right axis respectively.

❻ The SERIESPLOT statement has its own set of options that will govern how the plot will be displayed. For the diabetic patient profile plot, the series line along with the markers are to be displayed, so you need to use DISPLAY = ALL. The default display option is STANDARD, which does not display the markers.

❼ One of the options for the SERIESPLOT enables you to control the appearance of the markers: MARKERATTRS. You can control attributes such as color, size, and symbol. The marker is the symbol that is used to represent each data point on the series plot. As shown in Output 8-2, the markers are firebrick (red) filled circles.

❽ The LINEATTRS options enables you to control the appearance of the series line. You can control attributes such as line color, line pattern, and line thickness.

❾ Typically, a patient profile plot is generated for all patients in the data set. For illustration purposes, SAS Program 8-2 is only subsetting for the one patient in order to produce the plot in Output 8-2.

Output 8-2: Series Plot for Glucose (mg/dL) for Patient 0001

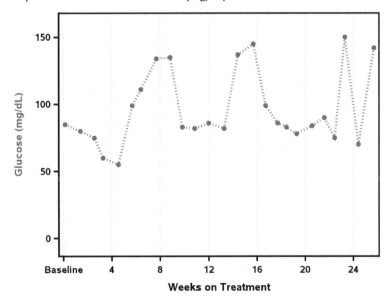

Note the same process used to create the Glucose series plot is repeated to produce the HbA1c series plot. Program 8-3 highlights the modifications needed to produce the HbA1c plot, which is displayed in Output 8-3.

Program 8-3: Template to Produce Series Plot for HbA1c (%)

```
/* formats to display time point */
proc format;
   picture dy2mo (round)      ❶
                0 = 'Baseline' (noedit)
        1 - high = 09 (mult = 0.035714);
run;

proc template;
   define statgraph hba1cprof;
      begingraph / border = false;
         layout overlay / xaxisopts = (label = 'Months on Treatment'  ❷
                               griddisplay = on
                               gridattrs = (pattern = 35 color=libgr)
                                labelattrs = (color = black weight = bold)
                                linearopts = (viewmin = 0 viewmax = 180
                                          tickvaluelist = (0 56 112 168)    ❸
                                          tickvalueformat = dy2mo.))

                          yaxisopts = (label = 'HbA1c (%)'  ❹
                              offsetmin = 0.07 offsetmax = 0.07
                              labelattrs = (color = cornflowerblue
                                         weight = bold)
                              linearopts = (viewmin = 0   viewmax = 8));

             seriesplot x = strtday y = hba1c / display = all   ❺
                 markerattrs = (symbol = squarefilled color = cornflowerblue
                             size = 6)
                 lineattrs = (color = cornflowerblue pattern = 12
                          thickness = 3);
         endlayout;
      endgraph;
   end;
run;

proc sgrender data = adam.diabprof template = hba1cprof;
   where usubjid = 'ABC-DEF-0001';
run;
```

❶ Similar to the format that displayed the time points in weeks, a format is created to display the time points for HbA1c in months.

❷ The options for the X axis are changed to account for the data being captured in months instead of weeks.

❸ Instead of using the TICKVALUESEQUENCE option that specifies a start and end of the tick marks sequence and how the tick marks should be incremented, a list of values for the tick marks can be provided with TICKVALUELIST. In addition, the new format is applied.

❹ The options for the Y -axis are also modified to account for the specific colors and minimum and maximum value for HbA1c.

❺ The options for the SERIESPLOT statement are also modified so that the series plot for HbA1c is shown in the color cornflower blue with filled squares as the markers and a line pattern of 12.

Output 8-3: Series Plot for HbA1c (%) for Patient 0001

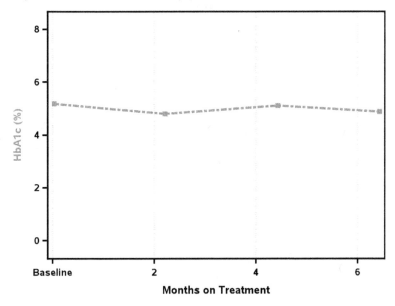

Because each series plot has its own set of axes and both plots need to reside on the same plot, you need to use the X2 and Y2 axes to allow for the display of the two lab tests. Program 8-4 illustrates the use of the X2 and Y2 axes with the results shown in Output 8-4.

Program 8-4: Template to Produce Series Plot for Both Glucose (mg/dL) and HbA1c (%) on the Same Graph

```
proc template;
   define statgraph ghprof;
      begingraph / border = false;
         layout overlay / xaxisopts = (label = 'Weeks on Treatment'
                           griddisplay = on
                           gridattrs = (pattern = 35 color=libgr)
                             labelattrs = (color = black weight = bold)
                              linearopts = (viewmin = 0    viewmax = 180
                                 tickvaluesequence = (start = 0 end = 168
                                                      increment = 28)
                                    tickvalueformat = dy2wk.))

                        yaxisopts = (label = 'Glucose (mg/dL)'
                            offsetmin = 0.07 offsetmax = 0.07
                            labelattrs = (color = firebrick
                                          weight = bold)
                            linearopts = (viewmin = 0 viewmax = 155))

                        x2axisopts = (label = 'Months on Treatment'  ❶
                            labelattrs = (color = black weight = bold)
                            linearopts = (viewmin = 0 viewmax = 180
                                 tickvaluelist = (0 56 112 168)
                                 tickvalueformat = dy2mo.))
```

```
                        y2axisopts = (label = 'HbA1c (%)' ❶
                                offsetmin = 0.07 offsetmax = 0.07
                                labelattrs = (color = cornflowerblue
                                                  weight = bold)
                                linearopts = (viewmin = 0  viewmax = 8));

            seriesplot x = strtday y = sgluc / display = all
                markerattrs = (symbol = circlefilled color = firebrick
                            size = 6)
                lineattrs = (color = firebrick pattern = 34 thickness = 3);

            seriesplot x = strtday y = hba1c / xaxis = x2 yaxis = y2  ❷
                display = all
                markerattrs = (symbol = squarefilled color = cornflowerblue
                            size = 6)
                lineattrs = (color = cornflowerblue pattern = 12
                            thickness = 3);
        endlayout;
      endgraph;
    end;
run;

proc sgrender data = adam.diabprof template = ghprof;
    where usubjid = 'ABC-DEF-0001';
run;
```

❶ The default axes used are the left and bottom axes. However, you can override the default behavior and use the right and top axes. When you use the right and top axes, you can control how they look using the same options that you would use for the left and bottom axes. The difference is that instead of XAXISOPTS and YAXISOPTS you would specify X2AXISOPTS and Y2AXISOPTS.

❷ If you want a plot to use the right and top axes, then you need to indicate that with the XAXIS = X2 and YAXIS = Y2 options on the plot statement. The default value is X and Y.

Output 8-4: Series Plot for Glucose (mg/dL) and HbA1c (%) on the Same Graph for Patient 0001

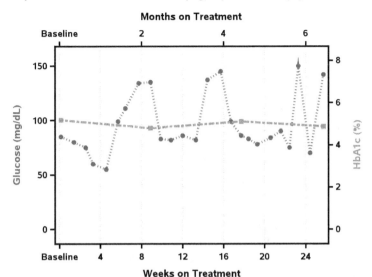

8.1.1.3 Creation of Scatter Plot for Hypoglycemic and Hyperglycemic Events

The next thing you need to create is the plot for the hypoglycemic events and hyperglycemic events. Upon first glance you might not think the plot is a scatter plot since typically a scatter plot has a scattering of points. However, this is indeed a scatter plot with the points scattered along the Y axis of each of the variables that represent the events. You can use the SCATTERPLOT statement for each event to produce the scatterplots. Program 8-5 illustrates the use of SCATTERPLOT to generate Output 8-5.

Program 8-5: Template to Produce Scatter Plot for Hypoglycemic and Hyperglycemic Events Over Time by Week

```
proc template;
   define statgraph aeprof;
      begingraph / border = false;
         layout overlay / xaxisopts = (label = 'Weeks on Treatment'
                              labelattrs = (color = black weight = bold)
                                linearopts = (viewmin = 0      viewmax = 180
                                  tickvaluesequence = (start = 0 end = 168
                                                      increment = 28)
                                                  tickvalueformat = dy2wk.))
                        yaxisopts = (linearopts = (integer = true));    ❶
            scatterplot x = strtday y = event1 /
               markerattrs = (symbol = diamondfilled color = red
                        size = 6);   ❷
            scatterplot x = strtday y = event2 /
               markerattrs = (symbol = trianglefilled color = orange
                        size = 6);   ❸
```

```
            endlayout;
         endgraph;
      end;
run;

proc sgrender data = adam.diabprof template = aeprof;
   where usubjid = 'ABC-DEF-0001';
run;
```

❶   If the axis values are evenly spaced integers, then you can use INTEGER = TRUE to indicate
    that only integer values should be displayed on the axis.

❷   For the hypoglycemic events (EVENT1), red filled diamonds are used to indicate when the
    event occurred.

❸   Hyperglycemic events (EVENT2) are indicated with orange filled triangles.

Note that it is possible to produce the scatter plot by events using one variable (for example,
EVENT) and one SCATTERPLOT statement with the GROUP = *variable*. However, this uses the
color and symbol attributes associated with GraphData#. Thus, in order to control the color and
symbols you would either need to split the data into two variables and specify the color and
symbol in each SCATTERPLOT statement as shown in Program 8-5. Or, you will control the color
and symbols using an alternate method such as modifying the individual components for
GraphData1 and GraphData2 via the styles using PROC TEMPLATE, which is beyond the scope of
this example.

**Output 8-5: Scatter Plot for Hypoglycemic and Hyperglycemic Events Over Time by Week for Patient
0001**

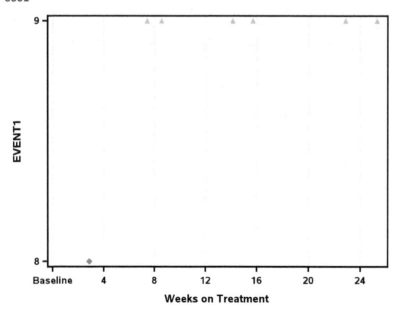

8.1.1.4 Creation of Highlow Plot for Duration of Medication by Type of Medication

Because you need to know the duration of each medication that was used to help control the patient's Glucose levels, a scatter plot is not sufficient. You need to use a plot that can span a period of time. In addition, since we need to incorporate the medications with the other plots, we need to have a numeric value to align with the X axis. As noted in Section 8.1.1.1 *Creation of Data Set with All Necessary Data to Build a Diabetic Patient Profile Plot* in Display 8-1, each medication was assigned a numeric code so that it could occupy the space below the series plot. In addition, each medication was assigned a value for CMCAP. If the medication had an end date, then CMCAP was set to SERIF, but if the medication was ongoing, then CMCAP was assigned FILLEDARROW. These values are used when creating the floating bars shown in Output 8-6 which is produced by Program 8-6.

Program 8-6: Template to Produce Highlow Plot for Duration of Medication by Type of Medication

```
proc template;
   define statgraph medprof;
      begingraph / border = false;
         layout overlay / xaxisopts = (label = 'Weeks on Treatment'
                          griddisplay = on
                          gridattrs = (pattern = 35 color=libgr)
                          labelattrs = (color = black weight = bold)
                          linearopts = (viewmin = 0    viewmax = 240   ❶
                               tickvaluesequence = (start = 0 end = 168
                                                    increment = 28)
                                          tickvalueformat = dy2wk.))

                    yaxisopts = (label = 'Medication'
                         offsetmin = 0.07 offsetmax = 0.07
                         linearopts = (viewmin = 1 viewmax = 7
                                       tickvaluelist = (1 3 5 7)));


         highlowplot y = med low = strtday high = endday / ❷
                                               group = acat     ❸
                 lineattrs = (thickness = 3)
                 type = line
                 lowcap = serif highcap = cmcap;     ❹

         endlayout;
      endgraph;
   end;
run;

proc sgrender data = adam.diabprof template = medprof;
   where usubjid = 'ABC-DEF-0001';
run;
```

❶ The X axis is extended beyond the original 180 days since a medication can be considered ongoing. Extending the X axis allows for an arrow to be displayed if the medication is ongoing.

❷ HIGHLOWPLOT produces either a vertical floating bar or a horizontal floating bar. If the Y option is specified, then it produces a horizontal floating bar and if X is specified it produces a vertical floating bar. Regardless if you use the X or Y option, you need to specify the LOW and HIGH options as well in order to indicate the minimum (LOW) and the maximum (HIGH) value for each categorical value.

❸ The GROUP option creates a set of floating bars for each group value, that is, each level of medication. Note that ACAT and MED variables have a one-to-one match. Thus, MED is used so that the floating bars can be placed in a specific order along the Y axis, which is linear. ACAT is used for the grouping so that the labels associated with each medication can be used if necessary. Recall that for the events that two separate variables were created in order to specify the different markers and colors for each event. With the medications, the GROUP option controls the colors for each highlow plot; therefore, only one variable is needed to distinguish between the different groups. By default SAS uses the colors associated with the style-elements GraphData1, GraphData2,…, GraphDataN for each unique group value.

❹ In the HIGHLOWPLOT statement, you can have LOWCAP or HIGHCAP to indicate how to display the start or end cap for each floating bar. LOWCAP and HIGHCAP options are not required. There are several types of caps:

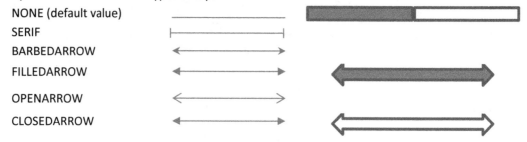

Output 8-6: Highlow Plot for Duration of Medication by Type of Medication for Patient 0001

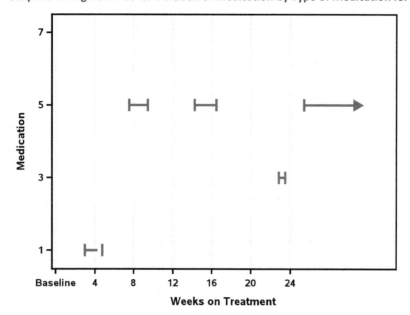

8.1.1.5 Creation of Diabetic Patient Profile Plot

Once you have determined how to create each individual component of the desired plot, you can start compiling the different pieces into one "master" template with a few minor additions. This "master" template combines all the necessary information into one output. Program 8-7 shows the final template that is used to produce the diabetic patient profile plot originally shown in Output 8-1 (reshown in Output 8-7).

Program 8-7: Template to Produce Diabetic Patient Profile Plot

```
proc template;
   define statgraph ptprof;
      begingraph / border = false;

      discreteattrmap name = "medcolor" / ignorecase = true  ❶
                                          trimleading = true;
         value 'Glucose Supplement' / lineattrs = (pattern = 1
                                                    color = green);
         value 'Elevating Agent'    / lineattrs = (pattern = 2
                                                    color = lightslategray);
         value 'Rapid-acting'       / lineattrs = (pattern = 35
                                                    color = firebrick);
         value 'Fast-acting'        / lineattrs = (pattern = 41
                                                    color = cornflowerblue);
      enddiscreteattrmap;
      discreteattrvar attrvar = colorvar var = acat  ❷
                      attrmap = "medcolor";

      layout overlay / xaxisopts = (label = 'Weeks on Treatment'
                         griddisplay = on
                         gridattrs = (pattern = 35 color=libgr)
                           labelattrs = (color = black weight = bold)
                             linearopts = (viewmin = 0    viewmax = 180
                               tickvaluesequence = (start = 0 end = 168
                                                    increment= 28)
                                   tickvalueformat = dy2wk.))
                       yaxisopts = (label = 'Glucose (mg/dL)'
                         offsetmin = 0.07 offsetmax = 0.07
                         labelattrs = (color = firebrick
                                       weight = bold)
                         linearopts = (viewmin = 0 viewmax = 155))
                       x2axisopts = (label = 'Months on Treatment'
                         labelattrs = (color = black weight = bold)
                         linearopts = (viewmin = 0 viewmax = 180
                                       tickvaluelist = (0 56 112 168)
                                       tickvalueformat = dy2mo.))
                       y2axisopts = (label = 'HbA1c (%)'
                         offsetmin = 0.07 offsetmax = 0.07
                         labelattrs = (color = cornflowerblue
                                       weight = bold)
                         linearopts = (viewmin = 0   viewmax = 8));

         seriesplot x = strtday y = sgluc / display = all
            markerattrs = (symbol = circlefilled color = firebrick
                           size = 6)
            lineattrs = (color = firebrick pattern = 34 thickness = 3);
```

```
                seriesplot x = strtday y = hba1c / xaxis = x2 yaxis = y2
                                                display = all
                    markerattrs = (symbol = squarefilled color = cornflowerblue
                                   size = 6)
                    lineattrs = (color = cornflowerblue pattern = 12
                                 thickness = 3);

                scatterplot x = strtday y = event1 /
                    markerattrs = (symbol = diamondfilled color = red size = 6)
                    name = 'hypo' legendlabel = 'Hypoglycemic Event';      ❹

                scatterplot x = strtday y = event2 /
                    markerattrs = (symbol = trianglefilled color = orange
                                   size = 6)
                    name = 'hyper' legendlabel = 'Hyperglycemic Event';     ❹

                highlowplot y = med low = strtday high = endday /
                                                    group = colorvar    ❸
                    lineattrs = (thickness = 3)
                    type = line lowcap = SERIF highcap = cmcap
                    name = 'med';    ❹

                discretelegend 'hypo' 'hyper' 'med' / location  = inside ❺
                                                      across = 1
                    autoalign = auto exclude = (" " ".")
                    valueattrs = (size = 6)   ❻;
            endlayout;
        endgraph;
    end;
run;

proc sgrender data = adam.diabprof template = ptprof;
    where usubjid = 'ABC-DEF-0001';
run;
```

❶ Sometimes you need to control the colors of specific values so that they are consistent from one patient to the next. Otherwise, you might end up with blue representing two or more different values across all the patients. There are several ways to control the color and one option is using an attribute map (Watson, 2020). An attribute map enables you to control not only the color but various other attributes as well, such as line patterns, markers, and size for each **formatted** data value that is specified. When defining your attribute map, if the data is formatted, then the formatted value must be used in the map definition. Section 2.5.5 *Producing Frequency Plot of Adverse Events Grouped by Maximum Grade for Each System Organ Class* explains the three components needed to create an attribute map using DISCRETEATTRMAP.

❷ After the attribute map has been defined, then it needs to be associated with the appropriate variable using the VAR option in the DISCRETEATTRVAR statement. In addition, this association needs to be provided with a name using the ATTRVAR option. The ATTRMAP option is used to indicate what attribute map is being associated with the variable indicated.

❸ The association name specified in the DISCRETEATTRVAR statement is then used in the corresponding plot statement.

❹ Because there are different data being captured in one plot, it is beneficial to identify what each piece represents. In order to label each component, a legend can be created. In order to allow a plot statement to be referenced in another statement within the template, the plot statement needs to be provided a name. In order to name the plot you would use the NAME option. In addition, you can choose to provide a label that is to be used in the legend plot by using LEGENDLABEL. If the legend is to use the values in the data, then LEGENDLABEL should not be provided.

❺ DISCRETELEGEND statement creates a legend based on the named plots indicated within the template. There are several options that you can use to control the location and display format of the legend. For example, you can have the legend displayed inside the graph or outside the graph with the LOCATION option. With the ACROSS option, you can indicate how many values should be displayed across in the legend. In addition, you can specify if the legend should be at the top, bottom (VALIGN option), left, right (HALIGN option) or allow SAS to determine the best location for the legend (AUTOALIGN option). Furthermore, you can have specific values suppressed from the legend by using the EXCLUDE option. For example, in the patient profile plot, missing values are suppressed because not all patients might take the same rescue medications.

❻ Similar to other GTL statements, you can control the color and font attributes for the values in the legend with VALUEATTRS option.

Output 8-7: Diabetic Patient Profile Plot for Patient 0001

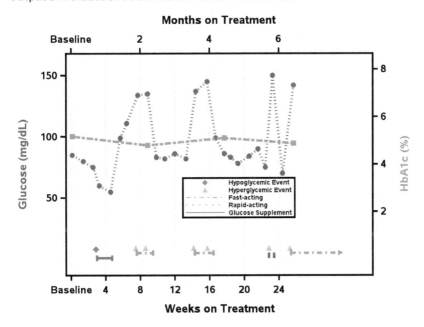

8.1.2 Using An Alternate Approach to Producing the Profile Plot

In Section 8.1.1 *Walking Through a Diabetic Patient Profile Plot*, the patient profile plot was created using the OVERLAY layout with all the plot statements within the same layout. However, if you wanted to split the lab data and event data and medication data into their own layouts, you can use a LATTICE layout as illustrated in Program 8-8. The output looks slightly different but shows a clear isolation of the lab results and the legend is placed in a consistent spot for all patients as shown in Output 8-8.

Program 8-8: Template to Produce Diabetic Patient Profile Plot Using an Alternate Approach

```
proc template;
   define statgraph ptprof2;
      begingraph / border = false;

         discreteattrmap name = "evtcolor" / ignorecase = true
                                             trimleading = true;    ❶
            value 'Hypoglycemia'  / markerattrs = (symbol = diamondfilled
                                                   color = red size = 6);
            value 'Hyperglycemia' / markerattrs = (symbol = trianglefilled
                                                   color = orange
                                                   size = 6);
         enddiscreteattrmap;
         discreteattrvar attrvar = colorevt var = aedecod
                     attrmap = "evtcolor";

         discreteattrmap name = "medcolor" / ignorecase = true
                                             trimleading = true;
            value 'Glucose Supplement' / lineattrs = (pattern = 1
                                                      color = green);
            value 'Elevating Agent'   / lineattrs = (pattern = 2
                                                      color = lightslategray);
            value 'Rapid-acting'      / lineattrs = (pattern = 35
                                                      color = firebrick);
            value 'Fast-acting'       / lineattrs = (pattern = 41
                                                      color = cornflowerblue);
         enddiscreteattrmap;
         discreteattrvar attrvar = colormed var = ACAT
                     attrmap = "medcolor";

         layout lattice / columns = 1 rows = 2    ❷
                        rowweights = (0.7 0.3);   ❸

            layout overlay /
                    xaxisopts = (display = (line)    ❹
                       griddisplay = on
                       gridattrs = (pattern = 35 color=libgr)
                       linearopts = (viewmin = 0 viewmax = 240
                          tickvaluesequence = (start = 0 end = 168
                                               increment = 28)))
                    yaxisopts = (label = 'Glucose (mg/dL)'
                       offsetmin = 0.07 offsetmax = 0.07
                       labelattrs = (color = firebrick weight = bold)
                       linearopts = (viewmin = 50 viewmax = 155))
                    x2axisopts = (label = 'Months on Treatment'
                       labelattrs = (color = black weight = bold)
                       linearopts = (viewmin = 0 viewmax = 240
                          tickvaluelist = (0 56 112 168)
                          tickvalueformat = dy2mo.))
                    y2axisopts = (label = 'HbA1c (%)'
                       offsetmin = 0.07 offsetmax = 0.07
                       labelattrs = (color = cornflowerblue
                                     weight = bold)
```

```
                          linearopts = (viewmin = 2 viewmax = 8));

                seriesplot x = strtday y = sgluc / display = all
                    markerattrs = (symbol = circlefilled color = firebrick
                                 size = 6)
                   lineattrs = (color = firebrick pattern = 34 thickness = 3);

                seriesplot x = strtday y = hba1c / xaxis = x2 yaxis = y2
                                                 display = all
                      markerattrs = (symbol = squarefilled
                                  color = cornflowerblue size = 6)
                    lineattrs = (color = cornflowerblue pattern = 12
                               thickness = 3);
              endlayout;

              layout overlay / walldisplay = none  ❺
                      xaxisopts = (label = 'Weeks on Treatment'  ❹
                         griddisplay = on
                         gridattrs = (pattern = 35 color=libgr)
                          labelattrs = (color = black weight = bold)
                          linearopts = (viewmin = 0 viewmax = 240
                             tickvaluesequence = (start = 0 end = 168
                                                  increment = 28)
                             tickvalueformat = dy2wk.))

                      yaxisopts = (display = (line)  ❻
                                  linearopts = (viewmin = 0 viewmax = 10))

                      y2axisopts = (display = (line)  ❻
                                  linearopts = (viewmin = 0 viewmax = 10));

                scatterplot x = strtday y = event / group = colorevt    ❼
                                                 name = 'evt'
                                                 yaxis = y2;

                highlowplot y = med low = strtday high = endday /
                                                 group = colormed
                    lineattrs = (thickness = 3)
                    type = line lowcap = SERIF highcap = cmcap name = 'med';
                endlayout;

               sidebar / align = bottom;    ❽
                  discretelegend 'evt' / title = 'Events:'
                     border = false exclude = (" " ".");
               endsidebar;
               sidebar / align = bottom;    ❽
                  discretelegend 'med' / title = 'Medications:'
                     border = false exclude = (" " ".");
               endsidebar;
            endlayout;
        endgraph;
     end;
run;

proc sgrender data = adam.diabmod template = ptprof2;
   where usubjid = 'ABC-DEF-0001';
run;
```

❶ As mentioned in Section 8.1.1.3 *Creation of Scatter Plot for Hypoglycemia and Hyperglycemia Events*, there are alternatives to controlling the color and symbols if you decide to use one variable with the GROUP option to distinguish between the type of events. If you use one variable to represent the events, then the data contained in EVENT1 and EVENT2 variables from Display 8-1 (Section 8.1.1.1 *Creation of Data Set with All Necessary Data to Build a Diabetic Patient Profile Plot*) would be consolidated into one variable, EVENT. Similar to how the colors were controlled for the highlow plot for the duration of medication, an attribute map for the events can be created that specifies the colors and symbols that are to be used for each event. With the duration of medication, you use the LINEATTRS option to control the attributes for the line. With the events, you use the MARKERATTRS to control the attributes of the markers.

❷ When using the LATTICE layout, you need to specify the number of columns and rows the graph output area will be broken into. For this example, there is only one column since you want both components to be displayed vertically. However, there are two rows. The first row is the series plots for the Glucose (mg/dL) and HbA1c (%) results. The second row represents the events and the medication durations.

❸ In addition to specifying the number of columns and rows, you also need to indicate how much space each piece will occupy in the graph output area. Because the series plots take up more space than the event markers and the medication duration bars, the series plot (row 1) is allotted 70% of the space while the event markers and the medication duration bars (row 2) are given 30%.

❹ The order in which the nested layouts are specified within the LATTICE layout is important because the cells of the lattice are populated by default as ROWMAJOR, that is, left to right; top to bottom. Since you want the series plot to take up the most space and it is listed first in the row weights option, then it needs to be the first layout that is specified. The layout for the first portion of the graph (row 1) can be copied from Program 8-4 in Section 8.1.1.2 *Creation of Series Plot for Glucose and HbA1c Lab Results* with a few modifications. Because you are using a LATTICE and you want both pieces to share the same X axis then the label and its attributes for the X axis are omitted from the layout. In addition, you need to indicate that DISPLAY = (LINE) so that only the line is displayed. Without this specified, then the X axis will have tick marks and a generic label. For the events and medication durations, the X-axis options need to contain all the information that pertains to both graphs. Therefore, it has the label and its corresponding attributes as well as the linear options.

❺ For the events and medication duration portions, the WALLDISPLAY = NONE option is used to prevent the plot's walls and outline from being displayed.

❻ The events and medication durations portions only need to have an axis line on the Y and Y2 axes so DISPLAY = (LINE) is used. In addition, the minimum and maximum value for the axes are specified in LINEAROPTS. Notice that both axes have the same options. This is so that either events or medication duration can use the Y axis while the other uses the Y2 axis. In order for both axes to display on the output both axes have to be used in one of the plot statements within the layout.

❼ In Section 8.1.1.4 *Creation of Highlow Plot for Duration of Medication by Type of Medication,* an attribute map was used to demonstrate how to control colors without having to create separate variables with each variable being used in a plot statement. In this alternative approach, you can use the same technique to control the color and symbols for the markers for the events.

❽ In order to have a legend, you can use SIDEBAR to create a discrete legend for each component of the graph. Note that only one GTL statement (for example, ENTRY or DISCRETELEGEND) or one layout can be specified within a SIDEBAR. Therefore, to have a legend for both the event and mediation duration, you need to have two side bars. You can specify the location of the SIDEBAR as top, bottom, left, or right.

Output 8-8: Diabetic Patient Profile Plot Using an Alternate Approach for Patient 0001

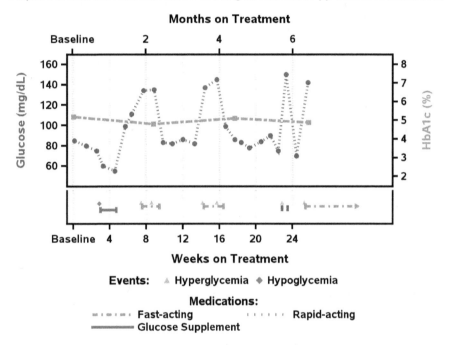

Instead of using the SIDEBAR statements to create the individual legends for events and medications, the GLOBALLEGEND statement can be used to create a legend that combines information from different plots. The template would be exactly the same as in Program 8-8 with the following exceptions:

● The SIDEBAR statements are removed.

● GLOBALLEGEND statement is added outside of the LATTICE layout.

Refer to Program 8-9 for illustration on how to use GLOBALLEGEND. The legend produced with GLOBALLEGEND will look different than the one produced using the DISCRETELEGEND in the SIDEBAR statement.

Program 8-9: Template to Produce Diabetic Patient Profile Plot Using an Alternate Approach with Global Legend

```
proc template;
    define statgraph ptprof3;
        begingraph / border = false;

            << discreteattrmap and discreteattrvar statements >>

            layout lattice / columns = 1 rows = 2
                             rowweights = (0.7 0.3);

                << nested layouts to produce patient profile plot >>
            endlayout;


            layout globallegend ❶ / border = false  ❷
                                  type = row   ❸
                                  legendtitleposition = top  ❹
                                  weights = preferred;   ❺
                discretelegend 'evt' /  across=1 title = 'Events:'
                                       border = false exclude = (" " ".");
                discretelegend 'med' / title = 'Medications:' across=1
                                       border = false exclude = (" " ".");
            endlayout;

        endgraph;
    end;
run;

proc sgrender data = adam.diabmod template = ptprof3;
    where usubjid = 'ABC-DEF-0001';
run;
```

❶   The GLOBALLEGEND layout enables you to create one legend. It consists of multiple discrete legends. Only one GLOBALLEGEND layout is allowed per template.

❷   As with all layouts, there are a variety of options that can be specified to control the look of the layout. BORDER option indicates whether you want a border around the global legend. Setting the option to FALSE removes the border. The default value is TRUE.

❸   The TYPE option specifies whether you want the global legend to be displayed as rows or as a column. Specifying ROW indicates that you want each legend to be displayed at the row level and legends appear side by side. Specifying COLUMN stacks all the legends so that they are vertical. The default value is ROW.

❹   You can choose whether you want the title to be displayed above the corresponding legend (TOP) or to the left of the corresponding legend. The default value is LEFT.

❺   If the TYPE = ROW, then it uses the WEIGHTS option to determine the width of each legend. You can choose to have the legends be the same width (UNIFORM), or you can specify specific weight values for each legend, or you can choose to have it based on the preferred width (that is, determined based on the values in the legend). This option is only used with TYPE = ROW and the default value is UNIFORM.

Comparing Output 8-8 with Output 8-9, you will notice that the legend in Output 8-8 has the labels for each symbol side by side and the title of the legend is to the left for Events but above for Medications. This is because SAS determines the best position for the title. However, the legend in Output 8-9 enables you to specify that you want the labels across the top.

Output 8-9: Diabetic Patient Profile Plot Using an Alternate Approach with Global Legend for Patient 0001

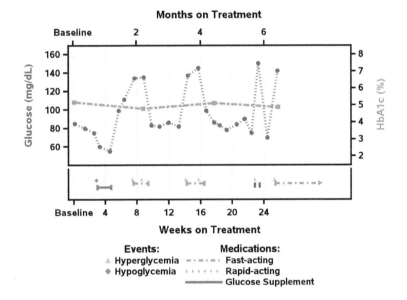

8.2 Chapter Summary

Although creating a patient profile graph might seem like an impossible task at first, it does not need to be daunting. Within GTL there are many types of plots with various types of options. By breaking down a seemingly impossible graph into individual components, you can create each piece of the graph using simple GTL statements. After you have built each piece, you can begin to incorporate each piece one at a time into a larger template and begin modifying the various options to fine-tune the graph to meet your desired needs. The best approach is typically starting with what you know and building from there.

Profile plots are typically created for multiple patients rather than for one patient. By using a BY statement within PROC SGRENDER, you can produce profile plots for multiple patients, and with the use of an attribute map, you can ensure that the color scheme is consistent from one patient to the next.

As you have seen there are different ways to produce the necessary output. However, the aesthetics of the graph is dependent on what the person who requested the output wants.

References

Watson, R. (2018). "Using a Picture Format to Create Visit Windows." Proceedings of the *SCSUG 2018* Conference. Austin, TX.

Watson, R. (2019). "Great Time to Learn GTL: A Step-by-Step Approach to Creating the Impossible." Proceedings of the *SAS Global Forum* 2019 Conference. Cary, NC: SAS Institute Inc. Dallas: SAS Institute.

Watson, R. (2020). "What's Your Favorite Color? Controlling the Appearance of a Graph." Proceedings of the *SAS Global Forum* 2020 Conference. Cary, NC: SAS Institute Inc.

Chapter 9: Selecting the Ideal Style, Output Format, and Graphical Options

9.1 Introduction to Styles, Output Format, and Graphical Options

Up through now, we have shown you how to create a variety of different graphs for the different data types. But these graphs are only as good as the aesthetic appearance of the output. Our goal when creating a graph is to effectively convey the correct message instead of hiding our results in a slew of data within a table or listing report or even within the graph itself. Part of making sure that our graph has the desired visual representation is making sure we are selecting the correct style, output format, and graphical options. If we do not select the optimal features, then we could "hide" a key point on our graph because either it was too small or the color blended too much with another. There could be various ways crucial information could inadvertently be diminished.

In this chapter, we will show you how to create a custom style to meet your needs. We will also go through some of the different types of ODS destinations and output formats. In addition, we will discuss how to select the ideal size and dots per inch (DPI) to get a crisp image. Lastly, we will give pointers on what type of output formats and ODS options should be used for specific types of documents.

9.2 Custom Styles

There are more than 50 different styles provided by SAS. While these styles have a wide range to meet most needs, there are times that you will need to control specific elements in order to produce the desired output. Most of the time companies have a specific color scheme or font that they want to use for all their outputs. In this situation, the company can create a style that is to be used. For example, for the development of this book, a custom style, CustomSapphire, is used so that all the outputs have a consistent look. Throughout the book we made modifications to graphs to meet the requirements for that particular graph. However, to avoid having to repeatedly make changes to elements on the plots statements, a custom style could be used.

In this section we touch briefly on how to create a custom style from an existing style, how to save the style, and how to reference the style.

9.2.1 Creating a Custom Style

When creating a custom style, it is typically easier to start with one of the 50 plus styles that SAS offers, rather than try to create one from scratch. Just to put things into perspective, the DEFAULT style, which is the parent style for a number of other SAS styles, has over 1000 lines of code. In order to see the code that is used to produce the specific SAS ODS Style, you can execute the code illustrated in Program 9-1. You can use the PROC TEMPLATE with the SOURCE statement to view the source code for both a permanent style or a temporary style. When referencing a permanent style you need to have "STYLES." preceding the style name.

Program 9-1: Viewing the Source Code for a Specific Style Template

```
%include "file-location\CustomSapphire.sas";

proc template;
    source CustomSapphire / expand; ❶
run;
```

❶ The EXPAND option shows the source code for any parent style that is used to produce the style being examined. For example, STYLES.SAPPHIRE is the parent for CUSTOMSAPPHIRE; STYLES.PEARL is the parent for STYLES.SAPPHIRE; STYLES.PRINTER is the parent for STYLES.PEARL; and STYLES.DEFAULT is the parent for STYLES.PRINTER. Thus, with the EXPAND option we can view the source code for all four parents in addition to the one we initially wanted to view. Without the EXPAND option, you would have to execute the PROC TEMPLATE with SOURCE statement 5 times.

There are a variety of elements that need to be defined when creating a style. Therefore, it is simpler to go through the many styles available from SAS and find the one that is closest to your needs and requires the modification of a few of the elements. There are style elements that govern how text appears in a graph (Table 9-1). With graph font elements, you can specify the fonts that can be used, the font size, the font weight, and the font style. There are also graph style elements that can be used to control how text appears as well as control the appearance of lines and markers. The difference between graph font elements and graph style elements is that there are additional attributes that can be specified to further customize the text. A list of common graph style elements along with attributes that can be modified are provided in Table

9-2. In addition, Display 9-1 illustrates some of these graph style elements as well as some graph style elements that are specific to box plots (Heath, 2018).

Table 9-1: Graph Font Elements

Graph Font Element	Graph Text That It Affects
GraphTitleFont	All titles
GraphFootnoteFont	All footnotes
GraphLabelFont	Axis labels and legend titles
GraphValueFont	Axis tick values and legend entries
GraphDataFont	Text where minimum size is necessary (such as for point labels and axis tables).
GraphUnicodeFont	Special glyphs (for example, α, ±, €) added to text in the graph
GraphAnnoFont	Text added as an annotation in the ODS Graphics Editor

Table 9-2: Common Graph Style Elements

Graph Style Element	Description	Attributes [1]
Graph	Size and the outer border appearance	Padding, BackgroundColor
GraphData*n*	Data attribute items associated with group *n*, where *n* is a value from 1 – 12	Color, ContrastColor, LineStyle, LineThickness, MarkerSymbol, MarkerSize
GraphDataDefault	Data attribute items associated with ungrouped data	Color, ContrastColor, LineStyle, LineThickness, MarkerSymbol, MarkerSize, StartColor, NeutralColor, EndColor
GraphGridLines	Line attributes for the horizontal and vertical grid lines for major tick marks	Color, ContrastColor, Displayopts, LineStyle, LineThickness

Graph Style Element	Description	Attributes[1]
GraphDataText [2]	Text attributes of point and line labels	Color, Font, FontSize, FontWeight, FontStyle
GraphValueText [2]	Text attributes associated with the tick values on the axis	Color, Font, FontSize, FontWeight, FontStyle
GraphLabelText [2]	Text attributes for the axis labels and legend title	Color, Font, FontSize, FontWeight, FontStyle
GraphFootnoteText [2]	Text attributes for the footnotes	Color, Font, FontSize, FontWeight, FontStyle
GraphTitleText [2]	Text attributes for the titles	Color, Font, FontSize, FontWeight, FontStyle
GraphWalls	Attributes for walls bounded by axes	BackgroundColor, Color, ContrastColor, Frameborder, LineStyle, LineThickness

[1] Color applies to filled areas. ContrastColor applies to markers and lines. StartColor, NeutralColor, and EndColor are used for color gradients.

[2] Font can reference the corresponding graph font element. For example, Font = GraphFonts("GraphDataFont") or you can overwrite the graph font element by specifying font, size, weight, and style on the style element. For example, fontfamily = "Times Roman" fontstyle = italic fontsize = 12pt.

Display 9-1: Illustration of Common Graph Style Elements

* Graph style elements specific to box plots.

9.2.1.1 Data Snippet of Subject Level Data Used to Illustrate Styles

Display 9-2 is a snippet of the data, which is used in Program 9-2. ADSL contains a record for each subject (USUBJID). It contains the treatment (TRT01P), age, and gender for each subject. In addition, summary statistics by treatment were generated and added to each record by treatment.

Display 9-2: Data Snippet of *Modified Analysis Subject Level Data Set* **(ADSLMOD)**

Obs	USUBJID	TRT01P	TRT01PN	AGE	N_C	MEAN_SD	MIN_MAX
1	01-701-1015	Placebo	0	63	4	66.0 (13.78)	52, 85
2	01-701-1023	Placebo	0	64	4	66.0 (13.78)	52, 85
3	01-701-1047	Placebo	0	85	4	66.0 (13.78)	52, 85
4	01-701-1118	Placebo	0	52	4	66.0 (13.78)	52, 85
5	01-701-1033	Xanomeline Low Dose	54	74	4	76.8 (7.18)	68, 84
6	01-701-1097	Xanomeline Low Dose	54	68	4	76.8 (7.18)	68, 84
7	01-701-1111	Xanomeline Low Dose	54	81	4	76.8 (7.18)	68, 84
8	01-701-1115	Xanomeline Low Dose	54	84	4	76.8 (7.18)	68, 84
9	01-701-1028	Xanomeline High Dose	81	71	2	74.0 (4.24)	71, 77
10	01-701-1034	Xanomeline High Dose	81	77	2	74.0 (4.24)	71, 77

9.2.1.2 CustomSapphire Style with No Modifications

To fully understand how the different elements affect different features in the graph, we need to have a comparison graph. Therefore, Program 9-2 produces two graphs of box plots, which are shown in Outputs 9-1 and 9-2.

Program 9-2: Box Plots Using CustomSapphire Style

```
ods rtf file = 'file-location\file-name.rtf' image_dpi = 300
        style = customSapphire;   ❶

proc sgplot data = adam.adslmod noautolegend; format trt01pn trt.;
   vbox age / category =  trt01pn extreme;
   xaxistable n_c  / x = trt01pn location = inside separator;
   xaxistable  mean_sd min_max / x = trt01pn location = inside;
   xaxis type = discrete label = "Treatment";
   yaxis type = linear label = "Change from Baseline";
run;

proc sgplot data = adam.adslmod; format trt01pn trt.;
   vbox age / group =  trt01pn extreme;
   xaxis type = discrete label = "Treatment";
   yaxis type = linear label = "Change from Baseline";
run;

ods rtf close;
```

❶ On an ODS statement, you can specify the STYLE that you would like to use. If STYLE is not specified, then the default style for the ODS destination and device will be used. Each device and ODS destination has default styles. For example, the default style for the RTF destination is the "RTF" style and for PDF destination the default style is "Pearl".

Output 9-1: Box Plot with Axis Table Using CustomSapphire Style

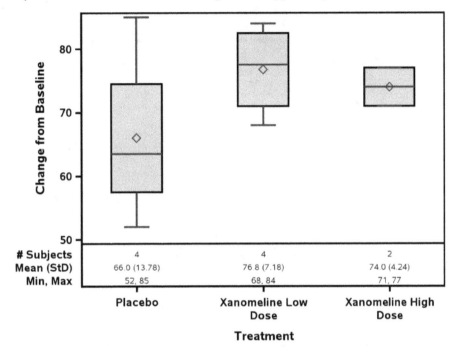

Output 9-2: Box Plot Grouped by Treatment Using CustomSapphire Style

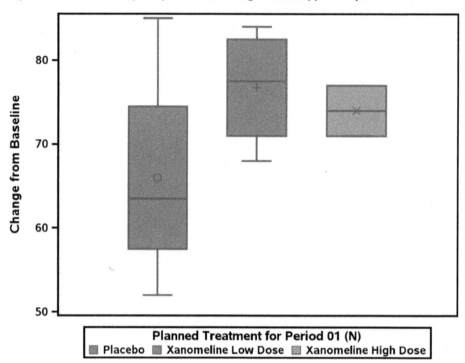

9.2.1.3 Modified CustomSapphire Style

Once you have identified the style that is closest to what you desire for your output you can begin to modify the features for that style using PROC TEMPLATE. Program 9-3 illustrates how the GRAPHDATAFONTS in the graph fonts element was modified to use the Courier New font with a size of 7. It further modifies the line thickness for the walls, outlines, whiskers on the box plot, and the median line on the box plot. In addition, the marker size is increased so that is not overshadowed by the increased width of the median line. Various text style elements are also adjusted.

Comparing Output 9-3 with Output 9-1, you will notice the difference between the original CustomSapphire style and the modified CustomSapphire Style. Output 9-3 is generated with Program 9-3 while Output 9-1 is generated with Program 9-2.

Program 9-3: Box Plot with Axis Table Using mySapphire Style, a Custom Style Built from CustomSapphire

```
proc template;
   define style mySapphire;      ❶
      parent = CustomSapphire;      ❷

      class GraphFonts / 'GraphDataFont' = ("Courier New", 7pt);      ❸

      class GraphWalls / lineThickness = 3px;      ❹
      class GraphOutlines / lineThickness = 3px;      ❹
      style GraphBoxWhisker from GraphBoxWhisker / lineThickness = 4px;      ❹
      style GraphBoxMedian from GraphBoxMedian / lineThickness = 4px;      ❹
      class GraphBoxMean / markersize = 10px;      ❹
      class GraphDataText / fontsize = 12pt      ❹
                           fontweight = bold
                           color = cadetblue;
      class GraphLabelText / fontfamily = "Times Roman"      ❹
                           fontstyle = italic
                           fontsize = 12pt;
      class GraphValueText / font = GraphFonts("GraphDataFont");      ❹
   end;
run;

ods rtf file = 'file-location\file-name.rtf' image_dpi = 300
      style = mySapphire;      ❺

proc sgplot data = adam.adslmod noautolegend;
   format trt01pn trt.;

   vbox age / category = trt01pn extreme;
   xaxis type = discrete label = "Treatment";
   yaxis type = linear label = "Change from Baseline";

   xaxistable n_c  / x = trt01pn location = inside separator;
   xaxistable  mean_sd min_max / x = trt01pn location = inside;
run;

ods rtf close;
```

❶ When creating a custom style, you need to give your new style a name.

❷ If you are starting from a previously defined style, then you need to indicate where your new style will inherit most of its features from. The parent style can be a SAS pre-defined style or another custom style. If it is a SAS pre-defined style, then you would indicate "STYLES." prior to the name of the style.

❸ Specify any font elements that need to be updated. When you specify the font element, you need to specify at a minimum the font family and the font size. According to SAS/GRAPH 9.4 documentation, GRAPHFONTS and GRAPHCOLORS are abstract elements because they are used to assign attributes for other elements defined within the style. (SAS Institute Inc., 2020)

❹ If there are any specific style elements that you want to modify the features, then you can do so with the CLASS statement. Another option is to use the STYLE statement with the FROM clause. Both CLASS statement and STYLE statement with FROM clause inherit from the parent style element. With the CLASS statement, the FROM is implied (Watson, 2020).

Notice that for GRAPHDATATEXT only the fontsize and fontweight are overwritten and the color is also specified. For GRAPHLABELTEXT, the font family is overwritten as well as the style and size but the weight is inherited from the parent style element. GRAPHVALUETEXT indicates that it should use the graph font element GRAPHDATAFONT instead of using the default GRAPHVALUEFONT.

❺ After you have created your custom style, you can use it to on any output that is produced by specifying it on the ODS statement using the STYLE option.

Output 9-3: Box Plot with Axis Table Using mySapphire Style, a Custom Style Built from CustomSapphire

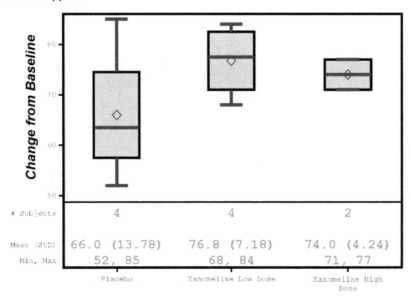

You can further enhance your custom style by modifying the GRAPHDATA*n* elements if you know you will have grouped data. Program 9-4 illustrates how GRAPHCOLORS and GRAPHDATA*n* elements can be adjusted to produce Output 9-4.

Program 9-4: Box Plot Grouped by Treatment Using mySapphire Style, a Custom Style Built from CustomSapphire

```
proc template;
   define style mySapphire2;
      parent = mySapphire;    ❶

      class GraphBoxMean / markersize = 14px;

      class GraphColors /   ❷
         'gcdata1' = CX90B328
         'gcdata2' = cornflowerblue
         'gcdata3' = CX9D2E14
         'gdata1' = darkred
         'gdata2' = CX90B328
         'gdata3' = cornflowerblue
         ;

      class GraphData1 / markersymbol = "circlefilled";    ❸
      class GraphData2 / markersymbol = "trianglefilled";    ❸
      class GraphData3 / markersymbol = "starfilled";    ❸
   end;
run;

ods rtf file = 'file-location\file-name.rtf' image_dpi = 300
      style = mySapphire2;

proc sgplot data = adam.adslmod noautolegend;
   format trt01pn trt.;

   vbox age / group = trt01pn extreme;
   xaxis type = discrete label = "Treatment";
   yaxis type = linear label = "Change from Baseline";
run;

ods rtf close;
```

❶ Similar to Program 9-3, the new style is inheriting from another custom style. This time we are building upon mySapphire, which was defined in Program 9-3.

❷ As mentioned in Program 9-3, GRAPHCOLORS is considered an abstract element because it is used by other elements within the style. So within mySapphire2, the colors defined for GCDATA1 – GCDATA3 and GDATA1 – GDATA3 in the GRAPHCOLORS statement are used by other elements that reference GCDATA1 – GCDATA3 and GDATA1 – GDATA3 within mySapphire2. Because we knew we only had three unique groupings, it was necessary to only modify the attributes for the first three groups. If there are more than three groups and only the first three groups were modified, then SAS would use the default color and contrastcolor inherited from the parent domain for the rest of the groups.

❸ By default GRAPHDATAn uses the colors defined with the style element GRAPHCOLORS; however, it is possible to overwrite the color as well as several other features such as markers. When specifying an attribute on the class statement for GRAPHDATAn, they must be defined sequentially with no gaps, starting with GRAPHDATA1. In other words, you cannot define a marker symbol for GRAPHDATA5 without defining a marker symbol for the previous GRAPHDATAn (that is GRAPHDATA1 – GRAPHDATA4). (Heath, 2018)

Output 9-4: Box Plot Grouped by Treatment Using mySapphire Style, a Custom Style Built from CustomSapphire

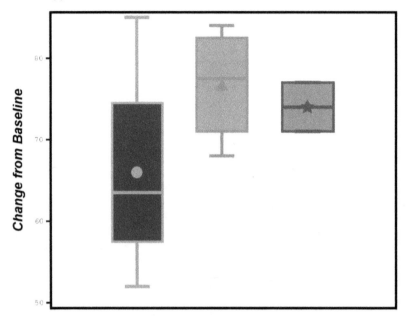

9.2.2 Saving the Custom Style Template

There might be times when you need to save a style to a permanent location so that it can be used by others within your organization. Program 9-5 demonstrates how to save the style for future use.

Program 9-5: Saving Custom Template to Permanent Template Store

```
libname adam "library path";

ods path ADAM.TEMPLAT (update) SASUSER.TEMPLAT (read) SASHELP.TMPLMST (read); ❶

proc template;
   define style mySapphire2;
      parent = mySapphire;
      class GraphBoxMean / markersize = 14px;

      class GraphColors /
         'gcdata1' = CX90B328
         'gcdata2' = cornflowerblue
         'gcdata3' = CX9D2E14
         'gdata1' = darkred
         'gdata2' = CX90B328
         'gdata3' = cornflowerblue;
      class GraphData1 / markersymbol = "circlefilled";
      class GraphData2 / markersymbol = "trianglefilled";
      class GraphData3 / markersymbol = "starfilled";
   end;
run;
```

❶ If your custom style is to be used by others, you can save your custom style to a permanent location. In order to save it, the ODS PATH statement needs to be updated so that it lists the location of where the new style should reside. SAS searches for the style based on the order in which the template stores are specified. The UPDATE option allows new templates to be written to that template store as well as allowing templates to be read from it. If you do not need to save to a permanent location and you want to use the default locations, then there is no need to specify the ODS PATH statement. SAS by default searches SASUSER.TEMPLAT then SASHELP.TMPLMST.

9.3 Output Delivery System Destination and Image Format

There are essentially two components to consider when determining the type of output you want to produce. You need to select which ODS destination you want the graph written to. You then need to select the output format type.

9.3.1 Selecting the Output Delivery System Destination

As mentioned in Section 9.2.1.2 *CustomSapphire Style with No Modifications*, each ODS destination has a default SAS style that is used. In order to achieve the desired look for your output, you need to know which destination you plan to use to produce the graph.

There are several destinations and each has its own default style. You are not required to use the default style associated with the destination. Table 9-3 lists the more common destinations used for creation of graphs along with the default style for each destination.

Table 9-3: Common ODS Destinations with Default and Recommended Styles

Destination Name	Description	Default Style
HTML	Opens, manages, or closes the HTML destination. The HTML destination produces HTML 4.0 output within the SAS windowing environment. The HTML 4.0 output can contain embedded style sheets.	HTMLBlue
PDF	Opens, manages, or closes the PDF destination. The PDF destination produces PDF output that is read by Adobe Acrobat and other applications.	Pearl
RTF	Opens, manages, or closes the RTF destination. The RTF destination produces output written in Rich Text Format that is used with Microsoft® Word.	RTF

Destination Name	Description	Default Style
POWERPOINT	Opens, manages, or closes the PowerPoint® destination. The PowerPoint destination produces PPTX output for use with Microsoft PowerPoint.	PowerPointLight
LISTING	Opens, manages, or closes the listing destination.	Listing

Program 9-6 illustrates how to specify the ODS destination and style needed for producing the desired output.

Program 9-6: Using ODS Statement to Write Output to Particular Destination

```
ods rtf  ❶
     file = 'file-location\file-output.rtf'  ❷
     style = mySapphire2;  ❸

<< SAS CODE >>

ods rtf close;  ❶


ods rtf  ❶
     file = 'file-location\file-output.rtf'  ❷
     style = mySapphire2;  ❸
ods pdf  ❶
     file = 'file-location\file-output.pdf'  ❷
     style = mySapphire2;  ❸

<< SAS CODE >>

ods _all_ close;  ❶
```

❶ ODS statement opens, manages and closes the indicated destination; RTF in Program 9-6. The first ODS statement opens the indicated destination so that any SAS DATA step and/or procedure output is directed to the appropriate destination, or destinations if multiple ODS statements are specified. In order to stop writing to the destination(s), you need to close it by using the CLOSE option on the last ODS statement after all the necessary SAS DATA step and/or procedure output has been produced. Essentially, any output that should be included in the destination file should be "sandwiched" between the first ODS statement that opens the destination and the last ODS statement that closes the destination. Note that you can write to multiple ODS destinations at the same time. In order to write to multiple destinations, you need to specify an ODS statement for each destination that you want to write to. After you have produced the output, you can use ODS _ALL_ CLOSE to close all open destinations or you can use ODS *destination* CLOSE for each individual destination. A word of caution when using ODS _ALL_CLOSE. When you specify ODS _ALL_ CLOSE this closes **all** open destinations, so if you have a destination open that you do not want to close, then you should probably use the individual ODS *destination* CLOSE statements instead.

❷   When you open an ODS destination, you have to indicate where you want to write the output to. This is done using the FILE option on the ODS statement.

❸   If you want to use a specific style, whether it is a SAS pre-defined or custom style, you can indicate which style to use with the STYLE option. If you do not specify the style that is to be used, SAS uses the default style for the indicated ODS destination.

Depending on which destination you select, there are a variety of options that can be used to control different features in the file as shown in Table 9-4.

Table 9-4: Common Options for HTML, PDF, and RTF ODS Destinations

Option	Description	Possible Values	HTML	PDF	RTF
BOOKMARKGEN	Controls generation of bookmarks	NO; YES		✓	
BOOKMARKLIST	Indicates whether bookmarks should be displayed If BOOKMARKGEN = NO, then BOOKMARKLIST defaults to NO	HIDE; NONE; SHOW		✓	
CONTENTS [1]	Controls whether a table of contents is generated		✓	✓	✓
DPI or IMAGE_DPI [2]	Specifies the image resolution		✓	✓	✓
COLUMNS	Specifies the number of columns to produce for each page	*n*		✓	✓
FILE [3]	Specifies the name of the output		✓	✓	✓
NOTOC or NOTOC_DATA	Removes the default table of contents Equivalent to using BOOKSMARKLIST = NONE and CONTENTS = NO			NOTOC	NOTOC_DATA

Option	Description	Possible Values	HTML	PDF	RTF
PATH	Specifies the location of an aggregate storage location	*aggregate-file-storage-location'*; *fileref*; *libref.catalog*; URL =	✓	✓	✓
PDFTOC	Controls the level which the table of contents is expanded	*n*		✓	
STARTPAGE	Controls page breaks in the file	NEVER; NO; NOW; YES; BYGROUP		✓	✓
STYLE	Specifies the style template that is to be used	*style-template*	✓	✓	✓
SGE	Generates a file that can be edited by ODS Graphics Editor	ON; YES; OFF; NO	✓		

[1] For HTML destination, an HTML file that contains the table of contents should be specified. For PDF destination, CONTENTS can have values of either YES or NO. For RTF destination, only the word CONTENTS needs to be specified.

[2] For HTML and RTF, the resolution option is IMAGE_DPI. For PDF, it is DPI.

[3] For PDF and RTF, the file name can include the path. For HTML, the path must be specified with the PATH option.

9.3.2 Selecting the Image Format

The ODS Graphics statement enables or disables the graphic processing. By default the ODS Graphics is enabled. You can also use a variety of options to set of the graphics environment. When creating a graph using ODS Graphics, several image characteristics are handled by default. Some of the different characteristics are image format, name, DPI and size. In this section we focus on the format. Section 9.4 *Options for Specific Types of Documents* discusses DPI and size.

Using ODS Graphics, you can specify the image format that is used to generate the image or the vector graphic. The image formats that are supported by SAS are shown in Table 9-5. However, not all image formats are supported for all ODS destinations. If you specify a format that is not supported by the destination, then SAS automatically defaults to the image format that is best suited for that destination. Program 9-7 illustrates how to specify the image format on the ODS Graphics statement.

Table 9-5: Image Formats Used to Generate Image Files

Image Format	Image Formats Supported in Vector Graphics Format
GIF	EMF [4]
JPEG (JPG)	EPS
PNG [1]	EPSI
SVG [2]	PCL
TIFF	PDF [3]
WMF	PS

[1] PNG is the default image format for EPUB, EPUB2, Excel, HTML, LISTING, PowerPoint, and PS destinations.

[2] SVG is the default image format for EPUB3 and HTML5 destinations.

[3] PDF is the default image format for PDF destination.

[4] EMF is the default image format for RTF and Measured RTF destinations.

Program 9-7: Specifying Image Format on ODS Graphics Statement

```
ods graphics on /      ❶
       outputfmt = file-type | STATIC;   ❷

<< SAS CODE >>

ods graphics off;     ❶
```

❶ ODS Graphics statement enables (ON or YES) or disables (OFF or NO) the graphic processing.

❷ One option that can be used on the ODS Graphics statement is OUTPUTFMT. You can specify the file type or you can specify the word STATIC. STATIC automatically determines the best format for the ODS destination.

For most of the destinations, the default image format is PNG. Typically, it is ideal to use the default image format unless there are special circumstances that require the use of an alternative format. PNG is an ideal format because it allows for transparency and supports a large number of colors. In addition, when writing the image to an ODS destination, a PNG format produces a smaller file size than using something like GIF.

However, there might be situations where you might need to use an image format other than the default. In some instances, you might be writing a graph to the PDF destination. The default value for OUTPUTFMT is PDF; however, you want to be able to extract the image from the PDF file so that it can be added to the report. Setting OUTPUTFMT = PNG will allow for the extraction of the image. Display 9-3 shows how the image will look in the PDF document if you click on the image. The image is highlighted in blue and will enable you to copy and paste the image to another document.

Display 9-3: PDF File with Extractable Image

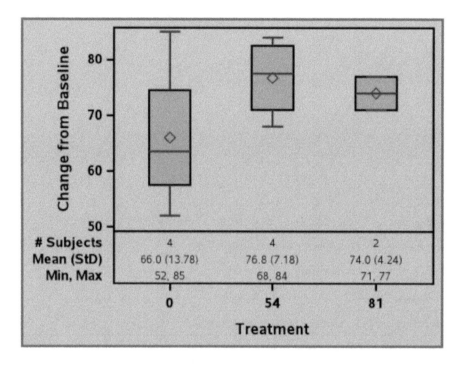

Another scenario in which the default image format might not be the best option is when a graph might be used for multiple reports and the writer of one report wants to make slight modifications to the existing graph, or a report writer might want to adjust the graph as they are developing the manuscript. The report writer could ask for the graph to be regenerated with the request, but in the clinical industry, this would require that the output goes through validation and review. If there are several iterations of changes, this could be a time-consuming process. If the changes are cosmetic, then it might be ideal to output the graph to multiple destinations with one of the destinations being the ODS HTML destination. When using the ODS HTML destination, you can indicate that the image format is set to SGE so that the output file can be opened in the ODS Graphics Editor allowing the post-processing of the graph to be done. With ODS Graphics Editor, the following items on the graph can be edited (SAS Institute Inc., 2020):

- Modify, add or remove titles and/or footnotes
- Modify the style, markers, line pattern, axis labels
- Add annotations such as arrows or circles to highlight a portion of the graph

9.4 Options for Specific Types of Documents

In this section, you will be shown various options and styles to consider applying for your output to be clear and to fit its intended purpose.

9.4.1 Setting Options for Documents Used for Presentations

Usually, when someone wants to produce graphs for a presentation, they either output a PNG file and insert that graph in their presentation, or they output an RTF file and copy the graph into their presentation. Generally, the graph that has been inserted or copied into the presentation has relatively small labels and tick values, and unless people are sitting in the front row when viewing the presentation, the values on the image can be difficult to read. One reason for the small fonts is because when the Statistician or Programmer are creating a PNG file, they generally use the default ODS GRAPHICS settings, and they use the default ODS LISTING settings. The default settings are fine, especially if your main purpose is to view the image with a graphics software program such as Paint or Windows Photo Viewer. However, if you are presenting the graphs then, as mentioned, it can be challenging to read from a distance. This section compares the default ODS settings to various settings that enhance the plot within your presentation. Program 9-8 shows you how to output a default listing plot.

Program 9-8: Creating Survival Plot with Default Settings

```
ods graphics / reset = all imagename = "Display9_4";   ❶
ods listing image_dpi = 96;   ❷

title1 'Product-Limit Survival Estimates';
title2 'With Number of Subjects at-Risk';

<< SAS CODE >>
```

❶ When the width and height are not specified, the ODS GRAPHICS options output the graph using the default width and height of 480px and 640px respectively.

❷ With the ODS LISTING destination, the default IMAGE_DPI is 96 for a PNG file. However, in your SAS configuration, the default IMAGE_DPI maybe 300. Using the default IMAGE_DPI of 96 with the default width of 480px and height of 640px produces an image that is outputted with the resolution 480px by 640px. The outputted resolution is the same as the default width and height because, all the default options were used. The resolution can be derived by first converting the size from pixels to inches. Thus, enabling you to equate 480px to 5 inches, and to equate 640px to 6.67 inches. The calculation used to convert the image size from pixels to inches is worked out by dividing the current pixels by the standard DPI. For example, 480 is divided by 96, and 640 is also divided by 96. The last piece to determine the resolution is to take the size in inches and multiple it by the image DPI that is used. So, if the width is 5 inches and there are 96 dots per inch (IMAGE_DPI = 96), the final resolution will have a width of 480px. The same formula can be applied to the height to get a final resolution of 640px. In this particular example, the pixels size that we started with yields the same numbers for the resolution. However, if you change the image DPI to 150 and use the same calculation, you can see that the size in pixels is not always equal to the resolution. Using an image DPI of 150, the width is calculated by using (480 / 96) * 150 = 750px.

Similarly, the height is calculated as (640 / 96) * 150 = 1000px. Thus, the resolution of the output image would be 750px * 1000px.

Display 9-4 shows how the default image looks when the image is inserted into PowerPoint and resized from 5 inches by 6.67 inches to 5.3 inches by 7.1 inches in order to maximize the plot space within the PowerPoint slide. You will notice that it could be a little challenging to read the labels, tick values, and the at-risk numbers.

Display 9-4: Survival Plot with Default Settings

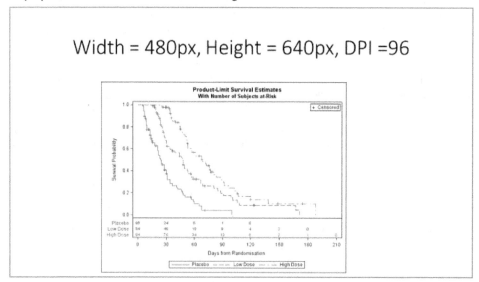

9.4.1.1 Survival Plot Using Default Dimensions and DPI of 300

In Display 9-4, you saw how the default settings could make it difficult to read the survival plot axis labels, tick values and at-risk numbers. Rendering the image with a DPI of 300 can make it easier to read the labels and values. The labels and values have a clearer resolution than the image that is produced with an IMAGE_DPI of 96. While the heights and widths of the files produced in Programs 9-8 (DPI of 96) and 9-9 (DPI of 300) are the same, the one produced with DPI of 300 has more pixels, which produce an image that makes the text easier to read. Program 9-9 shows you how to output an image with 300 DPI.

Program 9-9: Creating Survival Plot with Default Dimensions and DPI of 300

```
ods graphics / reset = all imagename = "Display9_5";  ❶
ods listing image_dpi = 300;  ❷

title1 'Product-Limit Survival Estimates';
title2 'With Number of Subjects at-Risk';

<< SAS CODE >>
```

❶ Similarly to Program 9-8, the ODS GRAPHICS options output the graph using the default width and height of 480px and 640px respectively because the width and the height are not specified.

❷ Using an IMAGE_DPI of 300 with a default width of 480px and a height of 640px produces an image that has the resolution of 1500px by 2000px. The width in the output is 1500px because the default width is 5 inches, and multiplying 5 inches by 300 DPI yields 1500px. Multiplying the default height, which is equivalent to 640 / 96, by 300 DPI yields 2000px.

Display 9-5 shows how the image looks when it is rendered with 300 DPI, has an output width of 640px and an output height of 480px, and is inserted into PowerPoint. You might notice that the fonts are slightly larger in Display 9-5 compared to Display 9-4; however, the fonts are still challenging to read.

Display 9-5: Survival Plot with Default Dimensions and DPI of 300

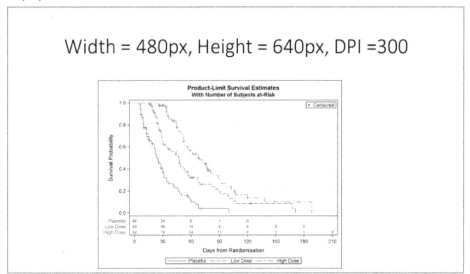

9.4.1.2 Survival Plot Using Smaller Dimensions and DPI of 300

In Displays 9-4 and 9-5, you saw that all the fonts in the graph are not that clear. It has a slightly grainy appearance. For clearer fonts, you can modify the LABELATTRS, TICKVALUEATTRS, and VALUEATTRS options in order to enlarge the text. However, you can use another method to increase the text size, and make the graph walls, graph lines and markers more prominent when viewing them in PowerPoint, and you can achieve this by resizing the graph to a smaller image while still using an image DPI of 300. The only drawback to resizing the image while using a DPI of 300 is that the fonts become too big for the graph in many cases. However, you will see how you can subsequently resize the text in order to obtain ideal proportions.

Resizing the image to a width of 3 inches and a height of 2.25 inches are good options to use. Program 9-10 shows you how you can resize the graph.

Program 9-10: Creating Survival Plot Using Smaller Dimensions and DPI of 300

```
ods graphics / reset = all imagename = "Display9_6"
            width=3in height=2.25in;   ❶
ods listing image_dpi = 300;   ❷

title1 'Product-Limit Survival Estimates';
title2 'With Number of Subjects at-Risk';

<< SAS CODE >>
```

❶ The WIDTH and HEIGHT options are used to resize the graph from 480px to 3 inches and 640px to 2.25 inches respectively.

❷ Using an IMAGE_DPI of 300 with a width of 3 inches and a height of 2.25 inches produces an output that has a resolution of 900px by 675px. The resolution can be derived by multiplying the number of inches by the DPI, that is, 3 inches * 300 DPI = 900px, and 2.25 inches * 300 DPI = 675px.

In Display 9-6, you can see that the titles, labels, tick values, and at-risk numbers are very prominent and stand out. Compared to Displays 9-4 and 9-5, the text is bolder and more comfortable to read, and the axis walls and border are bolder. However, the problem with Display 9-6 is that the text is too big. Because the text is so large, the Y axis label is not entirely showing, and the treatment legend spans over two lines. Also, because of the size of the text, the area of the survival curves has been reduced to accommodate the text. You will see how you can create a custom style to reduce the text and overcome these problems. You could also use the text attribute options to reduce the text, but using a custom style is recommended because you can easily apply that custom style to other graphs, and if you eventually like to make a change to any of the fonts, then you only have to change the attributes in one place.

Display 9-6: Survival Plot with Smaller Dimensions and DPI of 300

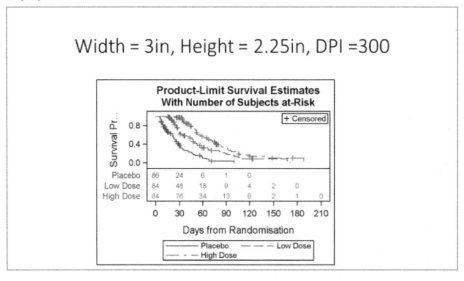

9.4.1.3 Survival Plot Using Smaller Dimensions, DPI of 300, and Custom Style

In Display 9-6, the graph stood out more than Displays 9-4 and 9-5. However, in some cases, the text was too big. The reason why the fonts are so large is because the fonts and other items within the graph are scaled using what is called a Scale Factor. The Scale Factor is determined using the following formula: (Render height / Design height) ** 0.25. The Render height is 2.25 inches, and the Design height is the default height, which is 5 inches. Thus, the Scale Factor is $0.45^{0.25}$ = 0.82. Therefore, even though the height of the graph has decreased by more than 50% from 5 inches to 2.25 inches, the font sizes have decreased less than 20%, so this makes some of the text on the plot too big. By default the SCALE option is turned on in ODS GRAPHICS. This can be turned off by using then NOSCALE option in the ODS GRAPHICS statement (Matange, 2012).

To overcome the problem of the text being too big, you can create a custom style to reduce the various text on the graph. Program 9-11 shows you how you can create and use a style, in conjunction with an image that is rendered with the same dimensions and DPI settings as Program 9-10, and create an output that is optimal for viewing in PowerPoint.

Program 9-11: Creating Survival Plot with Smaller Dimensions, DPI of 300, and Custom Style

```
proc template;
   define style styles.customstyle;   ❶
      parent = Styles.listing;   ❷
      style GraphFonts from GraphFonts /
         'GraphValueFont' = ("<sans-serif>, <MTsans-serif>",6pt)   ❸
         'GraphLabelFont' = ("<sans-serif>, <Mtsans-serif>",8pt)   ❹
         'GraphDataFont' = ("<sans-serif>, <Mtsans-serif>",6pt);   ❺
   end;
run;

ods graphics / reset = all imagename = "Display9_7"
            width = 3in height = 2.25in;   ❻
ods listing image_dpi = 300   ❼
         style = styles.customstyle;   ❽

title1 'Product-Limit Survival Estimates';
title2 'With Number of Subjects at-Risk';

<< SAS CODE >>
```

❶ The DEFINE STYLE statement creates a new style template called STYLES.CUSTOMSTYLE.

❷ The parent of the new style is STYLES.LISTING and STYLES.CUSTOMSTYLE will inherit all their style attributes from that parent style. Program 9-11 uses the LISTING parent style because that is the default style template for the ODS LISTING destination; thus for the purpose of illustration, it is reasonable to use that style template as the parent style. The parent style for most destinations can be any of the SAS styles available that closely meet your needs. However, there are some destinations where you should use specific styles or inherit from particular styles in order to yield satisfactory results, such as the POWERPOINT destination.

❸ As mentioned in Table 9-1, the GRAPHVALUEFONT helps to control the size of the axis tick values and legend entries. In STYLES.LISTING the default size for GRAPHVALUEFONT is 9pt. In the new style, STYLES.CUSTOMSTYLE, this has been reduced to 6pt, so that the tick values, censored legend entry, and treatment legend can be made smaller while still being easy to read.

❹ GRAPHLABELFONT helps to control the label sizes. In Display 9-6, the Y-axis label was too big and was truncated, so reducing the size of this attribute is ideal in this example. The default size is 10pt and this has been reduced to 8pt.

❺ GRAPHDATAFONT helps to control the size of the at-risk values within the table. The default size is 7pt and this has been reduced to 6pt.

❻ Similarly to Program 9-10, the WIDTH and HEIGHT options are used to resize the graph from 480px to 3 inches and 640px to 2.25 inches respectively. Recall that 480px by 640px is equivalent to 5 inches by 6.67 inches.

❼ Similarly to Program 9-10 the image is rendered with an IMAGE_DPI of 300 and with a width of 3 inches and a height of 2.25 inches. This produces an output that has a resolution of 900px by 675px.

❽ The STYLE option associates the output with the style template STYLES.CUSTOMSTYLE, and therefore reduces the font size, so it has the balance of being readable but not being too large.

Display 9-7 shows the survival plot that is rendered to a DPI of 300 and has an output size of 3 inches by 2.25 inches, and uses the custom style named STYLES.CUSTOMSTYLE. You will notice that it is easier to read the fonts on this display compared to Displays 9-4 and 9-5, and the fonts in the display are not too big like in Display 9-6.

Display 9-7: Survival Plot with Smaller Dimensions, DPI of 300, and Custom Style

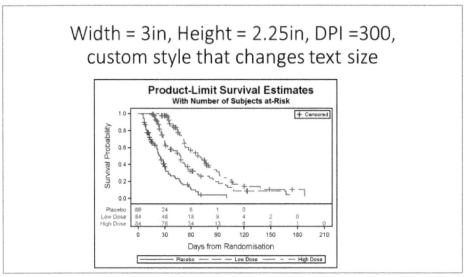

Display 9-8 shows a side-by-side comparison of Display 9-4 and Display 9-7 so that you can easily assess the difference, and notice that the fonts are indeed easier to read in Display 9-7 compared to Display 9-4. Not only are the fonts clearer, the Kaplan-Meier survival curves, axis wall, and border are darker and therefore stand out more. With this approach, it is easier to reduce the rendered height and rescale the fonts so that they are smaller. Otherwise, if you used a template with the default height and width and adjusted the font accordingly using an attribute option, you would also need to adjust the axis walls, borders, and graph lines with the different attribute options as well in order to produce a graph that looks similar to Display 9-7. In Section 9.4.1.4 *Using ODS PowerPoint to Output Graphs Directly into PowerPoint,* you learn how to achieve optimal settings by resizing all the necessary elements.

Display 9-8: Comparison of Default PNG Output and PNG Output with Optimal Settings

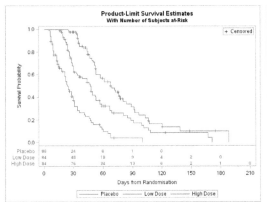
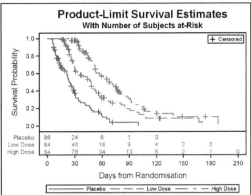

9.4.1.4   Using ODS PowerPoint to Output Graphs Directly into PowerPoint

You can also use ODS POWERPOINT to output your graphs directly into PowerPoint, and this could be beneficial if you need to include many graphs in your PowerPoint presentation. As shown in the paper "The Dynamic Duo: ODS Layout and the ODS Destination for PowerPoint" (Eslinger, 2016), it is also possible for SAS to create PowerPoint files, and those PowerPoint presentations can contain more than one graph per slide. This section focuses on how to create graphs for slides that present only one graph per slide.

Similar to creating PNG outputs, when creating PowerPoint presentations using custom settings instead of the default can result in a better-looking graph. Display 9-9 shows the result of using ODS PowerPoint with the default settings, and Display 9-10 shows the result when using more optimal settings. Display 9-11 shows you a side-by-side comparison of the two PowerPoint slides. Program 9-12 shows you how to create a PowerPoint file using the default settings. To create the PowerPoint files, GTL is used instead of SGPLOT. GTL is used because in GTL, you can use the ENTRYTITLE statement to add a title to the graph, that is, within the graph borders, and you are still able to use the TITLE statement to add the PowerPoint title. If you use SGPLOT, then you can only use the TITLE statement to produce either a PowerPoint title or a graph title.

Program 9-12: Creating PowerPoint File of SGRENDER output with Default Settings

```
goptions reset=all;
title 'Default PowerPoint Settings';  ❶
title2 ' ';  ❶

ods powerpoint file="&outputpath/Display9_9.pptx";  ❷

<< SAS CODE >>

ods powerpoint close;
```

❶ Using both the TITLE and TITLE2 statements simulates more closely the space that a standard PowerPoint presentation is assigned with. If only TITLE were used, then the space allocated for the title would be less than what the standard PowerPoint presentation allocates. Therefore, the TITLE2 was added to force a blank line and increase the title's assigned space.

❷ ODS POWERPOINT outputs a PPTX file containing the graph from the PROC SGRENDER output. If you use GTL with ODS POWERPOINT, then by default, the NOGTITLE option is used, and this results in a title that appears outside the graph borders, that is, in the PowerPoint title section. If you use SGPLOT with ODSPOWERPOINT, then by default, the titles appear inside the graph borders because the GTITLE option is switched on by default.

Display 9-9 shows the appearance of the SURVIVAL plot outputted with ODS POWERPOINT using the default settings. You will notice that the fonts, particularly the axis tick values and the at-risk numbers, are challenging to read.

Display 9-9: PowerPoint File of SGRENDER Output with Default Settings

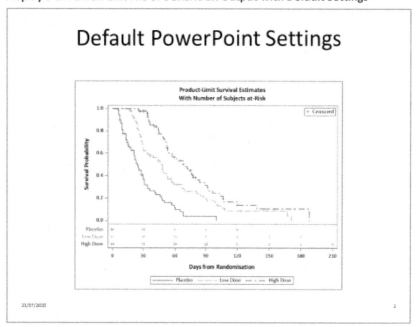

In Display 9-9, since the fonts are difficult to read, a method to increase the fonts and make them easier to read can be implemented. This method is the same method as what was done for the PNG outputs, namely, reducing the image size, and creating a style to lessen the huge unproportionate fonts further. However, a disadvantage of using this method is that you would have to manually resize the image in the PowerPoint file because it would be too small, and one of the reasons for using ODS POWERPOINT instead of the ODS LISTING method is to make the process of creating the right-size graphs in PowerPoint more automated, especially if you have lots of graphs to produce.

Another way to increase the fonts and make them easier to read is to create a style that has larger fonts. If you recall in Display 9-8, in addition to the fonts being larger and easier to see when using the optimal settings compared to the default settings, the Kaplan-Meier curves, axis walls, and axis border are also more prominent. You can also control these attributes in a style, and you will see how to do this in Program 9-13.

Program 9-13: Creating PowerPoint File of SGRENDER Output with Default Settings

```
proc template;
   define style styles.customstyle_pptx;   ❶
      parent = Styles.powerpointlight;   ❷
      style GraphFonts from GraphFonts /
         'GraphValueFont' = ("<sans-serif>, <MTsans-serif>",12pt)   ❸
         'GraphLabelFont'=("<sans-serif>, <MTsans-serif>",13pt,bold)   ❸
         'GraphDataFont'=("<sans-serif>, <MTsans-serif>",11pt)   ❸
         'GraphTitleFont' = ("<sans-serif>, <MTsans-serif>",14pt,bold);   ❸

      style GraphWalls from GraphWalls /
         linethickness = 2px;   ❹
      style GraphBorderLines from GraphBorderLines /
         linethickness = 2px;   ❺
      style GraphDataDefault from GraphDataDefault /
         linethickness = 3px   ❻
         markersize = 16px;   ❼
   end;
run;
ods powerpoint file="&outputpath/Output9_6.pptx" style=styles.customstyle_pptx;
   ❻

<< SAS CODE >>

ods powerpoint close;
```

❶ The DEFINE STYLE statement creates a new style template called STYLES.CUSTOMSTYLE_PPTX.

❷ The parent of the new style is STYLES.POWERPOINTLIGHT and STYLES.CUSTOMSTYLE_PPTX will inherit all their style attributes from that parent style. Program 9-13 uses the POWERPOINTLIGHT parent style because that is the default style template for the ODS POWERPOINT destination; thus, for the purpose of illustration it is reasonable to use that style template as the parent style. Another style created explicitly for the ODS POWERPOINT destination is called POWERPOINTDARK. If you inherit from a parent style other than POWERPOINTLIGHT or POWERPOINTDARK when using the POWERPOINT destination, such as the LISTING parent template, you run the risk of the output not turning out as expected. For example, if you were to use the LISTING parent style and you did not modify any of the title attributes, then the titles on the PowerPoint slide would be much smaller, compared to if you used the parent styles that were explicitly created for the POWERPOINT destination.

❸ The font sizes have been made more prominent than the default sizes to enable you to read the text on the graph easier.

- As mentioned in Table 9-1, the GRAPHVALUEFONT helps to control the size of the axis tick values and legend entries. In STYLES.POWERPOINTLIGHT, the default size for GRAPHVALUEFONT is 9pt. In the new style, STYLES.CUSTOMSTYLE the GRAPHVALUEFONT has been increased to 13pt so that the tick values, censored legend entry, and treatment legend can be made larger and read more comfortably.

- GRAPHLABELFONT helps to control the label sizes. The default size is 10pt and this has been increased to 13pt.

- GRAPHDATAFONT helps to control the size of the at-risk values within the table. The default size is 7pt and this has been increased to 11pt.

- GRAPHTITLEFONT helps to control the size of the titles. The default size is 11pt and this has been increased to 14pt.

❹ GRAPHWALLS helps to control the wall appearance. The wall is the box around the data where the axes are as well as the background behind the data area. The default line thickness is 1px and this has been increased to 2px.

❺ GRAPHBORDERLINES helps to control the border appearance around the graphs, layouts, and legends. The default line thickness is 1px and this has been increased to 2px.

❻ As mentioned in Heath (2018), GRAPHDATADEFAULT is the default style element for most non-grouped and no color response plots. The attributes GRAPHDATADEFAULT helps to control include MARKERSIZE, MARKERSYMBOL, LINETHICKNESS, LINESTYLE, CONTRASTCOLOR, and COLOR. The LINETHICKNESS attribute helps to control the appearance boldness of the lines for plots such as series plots and step plots. The default line thickness is 1px and this has been increased to 3px so that the Kaplan-Meier curves can be more easily seen.

❼ The MARKERSIZE attribute helps control the size of the markers, such as the size of the censored plus symbols on the Kaplan-Meier plot. The default marker size is 7px and this has been increased to 16px so that the markers can be easier to see.

❽ The STYLE option associates the output with the style template STYLES.CUSTOMSTYLE_PPTX, and therefore increases the font size, the graph wall thickness, the border thickness, the Kaplan-Meier curves, and the censored symbols. Thus, the PowerPoint graph is easier to read from a distance.

Display 9-10 shows the PowerPoint output with the custom settings. You can see that the fonts, Kaplan-Meier curve, titles, axis wall and border are more prominent compared to Display 9-9. You might be wondering why the default DPI was used instead of a DPI of 300, and this was because when using ODS POWERPOINT to output graphs, the DPI has a negligible effect. In fact, it can be argued that the graphs look slightly better when the default DPI settings are used. Using a higher DPI will only result in a larger PowerPoint file, and if you use a DPI that is too high, you could run into JAVA out of memory problems, that is, the following warning and error:

```
WARNING: A very large output size of (4000, 3000) is in effect. This could
make Java VM run out of memory and result in some Java exceptions. You
should reduce the output size or DPI settings.
ERROR: Java virtual machine exception. java.lang.OutOfMemoryError: Java
heap space.
```

Display 9-10: PowerPoint File of SGRENDER Using Custom Style

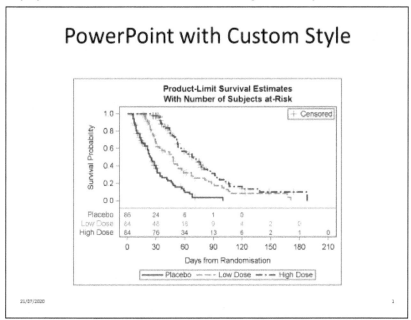

Display 9-11 shows a comparison of default PowerPoint output and PowerPoint output using a custom style. The graph on the left side is the default PowerPoint output, and the graph on the right side is the PowerPoint output using a custom style. You will notice that you can see the elements in the graph on the right side easier. Typically, when the graph is viewed in PowerPoint it is displayed on a projection screen, thus causing it to be bigger than how the graph is seen in the display. Therefore, when the graph is presented, people who are not sitting in the first row should also be able to see the graph elements comfortably.

Display 9-11: Comparison of Default PowerPoint Output and PowerPoint Output Using Custom Style

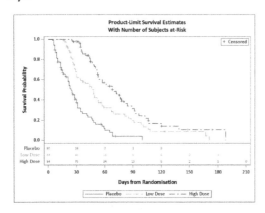 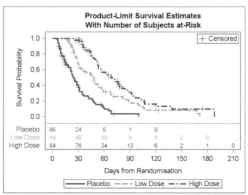

9.4.2 Setting Options for Documents Used for Publications

When producing graphs for publication, you are probably aware that some journals and publications have strict requirements, such as some publications only accept PDF, Encapsulated PostScript (EPS), and Tagged Image File Format (TIFF) images, while other publications accept only vector graphics. Sometimes the minimum DPI for image files must be 300. Some journals are very explicit about the items that they want to see on the graph. For example, if you are submitting a Kaplan-Meier plot to the Journal of Clinical Oncology, they require the patient's at-risk table be included with the Kaplan-Meier plot (Format My Manuscript, 2020).

The person requesting the graph might want the graph format created to specifications based on a graph that they have already seen accepted by the journal. Thus, it is important to be aware of all the requirements before you start to produce the graph and know who you are producing the graph for.

If you were to submit graphs within a paper to the journal Annals of Oncology, their requirements are below for any graphs not created in Microsoft Applications:

- EPS (or PDF): Vector drawings, embed all used fonts.

- TIFF (or JPEG): Color or gray-scale photographs (halftones), keep to a minimum of 300 DPI.

You can use SAS to produce EPS, PDF, TIFF, and JPEG files by merely changing the OUTFMT option within the ODS GRAPHICS STATEMENT. The PDF, TIFF, and JPEG file types all work with the ODS LISTING destination, and PS works with the ODS PS destination.

A Kaplan-Meier plot from the Annals of Oncology journal (Tabernero, 2018) is in Display 9-12. If you were asked to produce a plot like Display 9-12, would you know how to create it? In Chapter 5, you learned how to produce Kaplan-Meier curves. You also learned how to add statistics to the graph, such as the Hazard Ratio and Log-Rank *p*-value. In Program 5-6 of Section 5.2.3.3 *Generating Survival Plots for Subgroup Analysis*, you also learned how to create subgrouped

Kaplan-Meier plots. Thus, you can use that knowledge as a basis for creating a graph similar to Display 9-12. The primary difference in what was produced in Program 5-6 and Display 9-12 is the table of statistics are a little different because there are no left and right borders, and the appearance of the statistics is different because the Hazard Ratio is aligned in the middle of the treatment columns. However, to create the table of statistics like in the journal (Display 9-12) only requires the use of the DRAWLINE statement for the horizontal lines, and the DRAWTEXT statement in order to position the *p*-value in the middle of the columns. Program 9-14 shows you how to create the Kaplan-Meier plot in a similar way to the one provided in the journal.

Display 9-12: Example Kaplan-Meier Output from the Annals of Oncology Journal

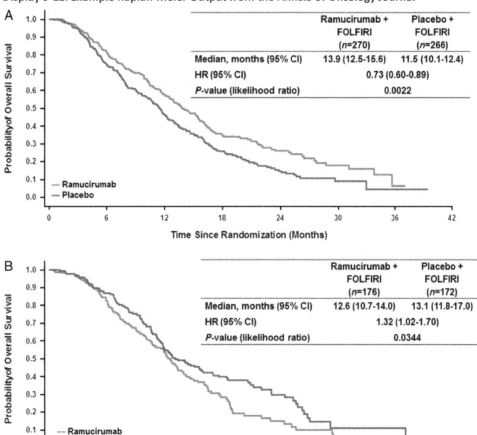

Program 9-14 shows you how to create a similar Kaplan-Meier plot as the plot in Display 9-12. However, the Kaplan-Meier plot is based on the overall data in the low and high dose groups instead of subgroups, as done in Display 9-12. Program 9-14 also shows you how to create a TIFF output at 300 DPI, as requested by many journals.

Program 9-14: Creating Similar Kaplan-Meier Plot for the Journal

```
proc format;
value trt
  0 = "Placebo"
  54 = "Low Dose"
  81 = "High Dose";

value $trt
  "0" = "Placebo"
  "54" = "Low Dose"
  "81" = "High Dose";
run;

ods output SurvivalPlot = SurvivalPlot;
ods output HomTests = HomTests(where = (test = "Log-Rank"));
ods output Quartiles = Quartiles;         ❶
ods output CensoredSummary = CensoredSummary;     ❷
proc lifetest data = adam.adtteeff plots = survival(atrisk = 0 to 210 by 30);
   where trtpn ne 0;
   time aval * cnsr(1);
   strata trtpn;
run;

ods output HazardRatios = HazardRatios;    ❸
proc phreg data = adam.adtteeff;
   where trtpn ne 0;
   class trtpn /ref = last;
   model aval * cnsr(1) = trtpn;
   hazardratio trtpn /diff = ref;
   format trtpn trt.;
run;

proc sql;
   select put(ProbChiSq, PVALUE6.4) into: log_rank_pvalue
   from HomTests;
quit;

proc sql;
   select put(hazardratio,4.2)||'('||put(waldlower,4.2)||',' ||
       put(waldupper,4.2)||')' into: HazardRatioCI   ❸
   from HazardRatios;
quit;

data quartiles2;
   set quartiles;
   where percent = 50;
   MedianVal = strip(put(Estimate, 2.));
   if MedianVal = "" then MedianVal = " NE";
   LowerLimVal = strip(put(LowerLimit, 2.));
   if LowerLimVal = '' then LowerLimVal = ' NE';
   UpperLimVal = strip(put(UpperLimit, 2.));
   if UpperLimVal = '' then UpperLimVal= ' NE';
   mediantext = cat(MedianVal, " (", LowerLimVal, "-", UpperLimVal, ")");   ❶
run;

proc sql;
  select mediantext into :mediantext1 - :mediantext2   ❶
  from quartiles2;
```

```
quit;

proc sql;
   select strip(put(total, best.)) into :totaltext1 - :totaltext2  ❷
   from Censoredsummary
   where trtpn in (54, 81);
quit;

proc template;
    define statgraph kmtemplate;

        mvar log_rank_pvalue HazardRatioCI Mediantext1 Mediantext2
            totaltext1 totaltext2;

        begingraph / datacontrastcolors =
                        (graphdata2:contrastcolor graphdata3:contrastcolor);  ❹
          layout overlay / xaxisopts = (label = "Time Since Randomisation (Days)"
            linearopts = (tickvaluesequence = (start = 0 end = 210 increment = 30)))
                        yaxisopts = (label="Probability of Overall Survival");

            stepplot x = time y = survival / group = stratum name = "Survival"
                legendlabel = "Survival";  ❺

            scatterplot x = time y = censored / markerattrs = (symbol = plus)
                group = stratum;

            discretelegend "Survival" / location = inside  ❺
                                        autoalign = (bottomleft)
                                        across = 1 down = 2;

          drawline x1 = 29.0 y1 = 0.98 x2 = 99.0 y2 = 0.98 /  ❻
             x1space = datapercent  y1space = datavalue
             x2space = datapercent  y2space = datavalue
             lineattrs = (pattern = 1 color = black thickness = 1px);

          drawtext textattrs = (size = 8) "Low" / justify = center ❼
             x = 70.0 y = 0.95 xspace = datapercent yspace = datavalue width = 30;
          drawtext textattrs =(size = 8) "High" / justify = center ❼
             x = 92.0 y = 0.95 xspace = datapercent yspace = datavalue width = 30;

          drawtext textattrs =(size = 8) "Dose" / justify = center ❼
             x = 70.0 y = 0.90 xspace = datapercent yspace = datavalue width = 30;
          drawtext textattrs =(size = 8) "Dose" / justify = center ❼
             x = 92.0 y = 0.90 xspace = datapercent yspace = datavalue width = 30;

          drawtext textattrs =(size = 8) "n=(" totaltext1 ")" / justify = center❼
             x = 70.0 y = 0.85 xspace = datapercent yspace = datavalue width = 30;
           drawtext textattrs =(size = 8) "n=(" Totaltext2 ")" / justify = center❼
             x = 92.0 y = 0.85 xspace = datapercent yspace = datavalue width = 30;

          drawline x1 = 29.0 y1 = 0.825 x2 = 99.0 y2 = 0.825 /  ❻
             x1space = datapercent y1space = datavalue
             x2space = datapercent y2space = datavalue
             lineattrs = (pattern = 1 color = black thickness = 1px);

          drawtext textattrs =(size =8) "Median, months (95% CI)" / anchor = left❽
             justify = left x =30.0 y =0.80 xspace = datapercent yspace = datavalue
             width = 45;
```

```
                drawtext textattrs =(size = 8)  Mediantext2 / justify = center x = 70.0❶
                    y = 0.80 xspace = datapercent yspace = datavalue width = 30;
                drawtext textattrs =(size = 8)  Mediantext1 / justify = center x = 92.0❶
                    y = 0.80 xspace = datapercent yspace = datavalue width = 30;

                drawtext textattrs =(size =8) "HR (95% CI)" / anchor=left justify=left❸
                 x = 30.0 y = 0.75 xspace = datapercent yspace = datavalue width = 45;
                drawtext textattrs =(size = 8) hazardratioci / justify = center ❸
                   x = 81 y = 0.75 xspace = datapercent yspace = datavalue width = 30;

                drawtext textattrs=(size = 8 style = italic) ❸
                   "P" textattrs =(size = 8) "-value (likelihood ratio)" / anchor = left
                   justify = left x =30.0 y =0.70 xspace = datapercent yspace = datavalue
                   width = 45;
                drawtext textattrs =(size =8) log_rank_pvalue / justify = center ❸
                   x = 81 y = 0.70 xspace = datapercent yspace = datavalue width = 30;

                drawline x1 = 29.0 y1 = 0.675 x2 = 99.0 y2 = 0.675 /    ❻
                    x1space = datapercent y1space = datavalue
                    x2space = datapercent y2space = datavalue
                    lineattrs = (pattern = 1 color = black thickness = 1px);
             endlayout;
          endgraph;
      end;
   run;

   ods listing image_dpi = 300;   ❾
   ods graphics / reset = all imagename = "Output9_5" outputfmt = tiff;   ❿

   title1 'Cumulative Incidence';
   title2 'With Number of Subjects at-Risk';

   proc sgrender data = Survivalplot template = kmtemplate;
      format stratum $trt.;
   run;
```

❶ Using the ODS statement, PROC LIFETEST outputs the QUARTILES table. This table contains the median survival times and the confidence limits for the low dose and high dose treatment groups. The median, lower, and upper confidence intervals are concatenated and stored in macro variables, and are used in GTL and displayed in the Kaplan-Meier output.

❷ PROC LIFETEST outputs the CENSOREDSUMMARY table using the ODS statement. The CENSOREDSUMMARY table contains the number of patients for the low dose and high dose treatment groups. PROC SQL stores the number of patients in each treatment group in macro variables, and GTL references the macro variables and displays the number of patients in the Kaplan-Meier output.

❸ PROC PHREG calculates and outputs the hazard ratios and the 95% confidence intervals in the HAZARDRATIOS table. The HAZARDRATIOS table contains the hazard ratios of the high dose against the low dose. PROC SQL combines the hazard ratio with the 95% confidence interval and stores in a macro variable for use within the GTL statements to display on the Kaplan-Meier output.

❹ Only the low dose and high dose treatments are present in the SURVIVALPLOT data. The Placebo data has been removed in order to closely mimic the journal figure. However, in order to keep the low and high dose the same color as the other plots, the following options were used in the BEGINGRAPH statement: DATACONTRASTCOLORS = (GRAPHDATA2:CONTRASTCOLOR GRAPHDATA3:CONTRASTCOLOR). Where GRAPHDATA2:CONTRASTCOLOR corresponds to the low dose and GRAPHDATA3:CONTRASTCOLOR corresponds to the high dose. It is not necessary to use GRAPHDATAn:CONTRASTCOLOR to specify the colors, you can use any one of the seven valid color-naming schemes to specify the desired colors. For example, you could have specified DATACONTRASTCOLORS = (CXA23A2E CX01665E) or DATACONTRASTCOLORS = (FIREBRICK TEAL) (Watson, 2020).

❺ The STEPPLOT statement uses the NAME to provide a name to the statement so that it can be referenced in other GTL statements within the template. In addition, it uses the LEGENDLABEL option so that when the legend is displayed it uses the value indicated. The LOCATION = INSIDE, AUTOALIGN = (BOTTOMLEFT), ACROSS = 1 and DOWN = 2 options are used in the DISCRETELEGEND statement to position the treatment legend in the bottom left corner, similarly to the position of the legend in Display 9-12.

❻ The DRAWLINE statements draw the horizontal lines above the column labels as well as above and below the median survival times, hazard ratios, and *p*-values. The X1 and Y1 are arguments are where the line starts and the X2 and Y2 arguments are where the line ends. For X1SPACE and X2SPACE the DATAPERCENT option is used. Therefore, values in X1 and X2 will be treated as a percentage of the X axis. Hence, if the X1 value is 50, the line will start in the middle of the X axis. For Y1SPACE and Y2SPACE the DATAVALUE is used. So Y1 and Y2 values will start at values with the same coordinates as the values on the Y axis.

❼ The first six DRAWTEXT statements add the treatment names and patient counts to the plot. You will notice that the low dose treatment label is centered at 70% of the X axis and high dose treatment label is centered at 92% of the X axis. The column labels are split across three lines. Thus for "Low" and "High" the texts are at the data value of Y = 0.95; "Dose" text is at the data value of Y = 0.90 and the patient counts are at the data value of Y = 0.85. Because each text is displayed at a different X and Y point, four DRAWTEXT statements are needed to display column labels in the embedded table.

❽ The last seven DRAWTEXT statements add the desired statistics as well as a label for each of the statistics. The median and 95% confidence interval are displayed under each dose column; thus, X corresponds to 70 and 92 so that the values are centered under the appropriate column header. Since the hazard ratio and log-rank *p*-value are to be plotted right in the middle of the low dose and high dose, the X option is set to 81 in order to position the hazard ratio and log-rank *p*-value between the two-column headers because the low dose and high dose used the X = 70 and X = 92 options respectively. The DRAWTEXT statements use the macro variables that are specified in the MVAR statement to populate the table.

❾ The IMAGE_DPI = 300 option renders the output with a DPI of 300.

❿ The OUTPUTFMT = TIFF option produces a TIFF output. Output 9-5 shows the Kaplan-Meier plot is similar to the journal's plot. It is important to note that depending on the platform and configuration of your version of SAS, you might have destinations open that do not support TIFF images. You might see a warning similar to this:

```
WARNING: SASREPORT13 destination does not support TIFF images. Using the
default static format.
```

To avoid this warning, you can close all the destinations by using ODS _ALL_ CLOSE and then open only the destination that you need to output your graph to.

Output 9-5: Kaplan-Meier Output Similar to Annals of Oncology Journal's Plot

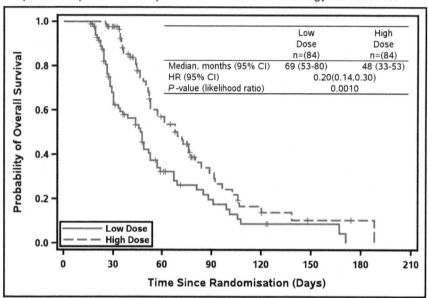

9.4.3 Setting Options for Documents Used for Medical Writer

Medical writers help pharmaceutical companies and contract resource organizations compile well-structured clinical trial results that clearly and concisely present information. Medical writers produce the results by extracting outputs created by the programmers and putting them into a report that ties the protocol, data, data exceptions/anomalies, and results together. If you are producing a graph for medical writers, sometimes you can go back and forth regarding the appearance of the graph. In similar situations where you are going back and forth regarding the appearance of the graph, or you know that the customer would like to edit the graph, then producing an editable graph could be a viable solution. Although, the receiver would need to have the ODS Graphics Editor to edit the graph.

You can create an editable graph by using the SGE = ON option in your ODS destination, and SAS will output the SGE figure with the original figures that you were creating. Program 9-15 gives an example of creating an editable file using SGE file.

Program 9-15: Creating Editable File with SGE

```
title1 'Cumulative Incidence';
title2 'With Number of Subjects at-Risk';

ods listing image_dpi = 300 sge = on;   ❶

<< SAS CODE >>
```

❶ The SGE = ON option produces an editable SGE file that can be edited with the ODS Graphics Editor.

You can open the editable file with ODS Graphics Editor, and edit titles, footnotes, labels, legend entries, and text on the graph. You can also change the colors and thickness of the lines. Display 9-13 shows an illustration of using the ODS Graphics Editor. Once you have opened the graph in ODS Graphics Editor, you can modify several features of the graph by opening the plot properties. To display the plot properties, you can right-click on the graph. Once you have opened the properties, you can modify the background and wall area as shown in Display 9-13. If the Fill box is checked you can select the color needed or you can allow SAS to automatically determine the color. In addition, you can decide whether you need to remove the outlines corresponding to the background and wall. If you uncheck the outline box under the Background section, the border around the graph area is no longer displayed and if you uncheck the outline box of the Wall section, you will notice that the top and right border of the wall area are removed as shown in Display 9-14.

Display 9-13: ODS Graphics Editor Illustration of Background and Wall Fill Area Modifications

Display 9-14: ODS Graphics Editor Illustration with Background and Wall Outline Removed

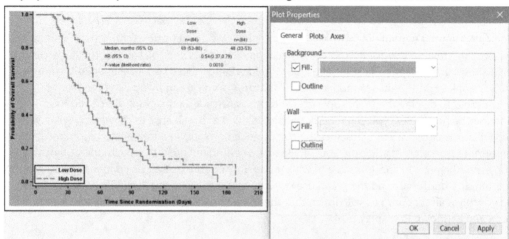

Not only can you modify the fill area and the outlines of the graph, you can modify various aspects of the graph itself. You can move the legend to a different location within the wall are or outside the wall area as shown in Display 9-15. You can also modify the text on the graph by clicking the text as demonstrated in Display 9-16. Notice that in Display 9-16 that the column headers in the embedded table are modified to include the treatment name, as well as the color of the headers are changed to correspond to the Kaplan-Meier curve.

Display 9-15: ODS Graphics Editor Illustration of Moving Legend to a Different Location

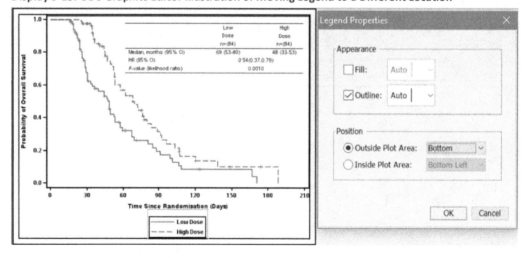

Display 9-16: ODS Graphics Editor Illustration of Modifying Text

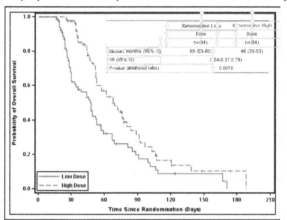

9.5 Options for Different Countries

If you need to produce graphs for customers in Europe or South America, you might be asked to provide graphs that display decimal commas instead of decimal places on the tick values. To display decimal commas instead of decimal places within your variables, or macro variables, you can use the SAS format NLNUM in conjunction with the LOCALE system option. When NLNUM and the LOCALE system option are used together, the numeric format is displayed according to the specified locale. So, if the specified locale uses a decimal comma, such as German and Spanish locales, then decimal commas are displayed.

You can use the TICKVALUEFORMAT option to format the tick values on a graph. In Display 9-16, to make the Probability of Overall Survival, display decimal commas instead of decimal places for Germany, you can use the TICKVALUEFORMAT = NLNUM3.1 option in conjunction with the LOCALE system option and specify the GERMAN_GERMANY locale. As previously mentioned, to display decimal commas for calculated statistics, such as the Hazard Ratio, you can use the NLNUM format when creating the macro variable in order to display the decimal commas. For displaying *p*-values, such as the log-rank *p*-value, you can use the NLPVALUE SAS format, to display decimal commas. Program 9-16 shows you an example program of creating tick values and macro variables with decimal commas:

Program 9-16: Creating Tick Values and Macro Variables with Decimal Commas

```
<< SAS CODE for Creating Dataset and Macro Variables >>

options locale = German_Germany;   ❶

proc template;
   define statgraph kmtemplate;
      mvar log_rank_pvalue   ❷
           HazardRatioCI Mediantext1 Mediantext2 totaltext1 totaltext2;   ❸
      begingraph / datacontrastcolors = (graphdata2:contrastcolor
                                         graphdata3:contrastcolor);
         layout overlay /
            xaxisopts = (label = "Time Since Randomisation (Days)"
```

```
             linearopts = (tickvaluesequence =
                                (start = 0 end = 210 increment = 30)))
         yaxisopts = (label="Probability of Overall Survival"
             linearopts = (tickvalueformat = nlnum3.1));   ❹


     << SAS CODE for rest of the graph as seen in Program 9-14 >>


         endlayout;
      endgraph;
    end;
run;

ods listing image_dpi = 300;

title1 'Cumulative Incidence';
title2 'With Number of Subjects at-Risk';

proc sgrender data = Survivalplot template = kmtemplate;
   format stratum $trt.;
run;
```

❶ The LOCALE system option is used to specify the locale, which reflects the local conventions, language, and culture of a geographical region (SAS Institute Inc., 2020). The default locale value is ENGLISH_UNITEDSTATES. For this example, the GERMAN_GERMANY locale value is used. The LOCALE system option is case insensitive.

❷ In Program 9-14, you see that the *LOG_RANK_PVALUE* macro variable is created with the PVALUE6.4 format. With the requirement of decimal commas for Germany, you can change the format to NLPVALUE6.2 instead. Using this format, with the GERMAN_GERMANY LOCALE system option, displays decimal commas in the *p*-value.

❸ Similar to the *LOG_RANK_PVALUE* macro variable, the *HAZARDRATIOCI, MEDIANTEXT1, MEDIANTEXT2, TOTALTEXT1,* and *TOTALTEXT2* macro variables, can be created with the NLNUM4.2 format instead of the 4.2 format to allow for the required decimal commas. Using this format, with the GERMAN_GERMANY LOCALE system options, display decimal commas in the statistics.

❹ The YAXISOPTS = (LINEAROPTS = (TICKVALUEFORMAT = NLNUM3.1)) displays decimal commas on the Y-axis tick values.

Output 9-6 shows an example Kaplan-Meier plot using decimal commas instead of decimals. You can see that the Y axis has decimal commas, and the Hazard Ratio and Confidence Intervals around the Hazard Ratios use decimal commas. In addition, you can see that the *p*-value is using decimal commas.

Output 9-6: Kaplan-Meier Output Using Decimal Commas

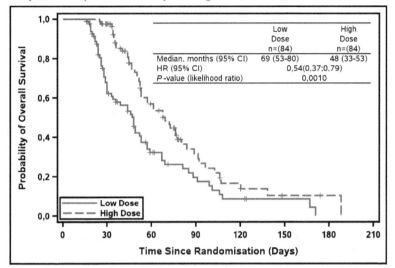

9.6 Chapter Summary

With style template, you can change the appearance of the items on your graph. You can change the font size of the axes labels, and the tick values, and you can change the size of the markers, and change many more attributes. If you are producing many graphs and want the graphs to have a consistent look, then using style templates can be quicker and easier to use than changing the relevant SGPLOT or GTL attribute options.

If you want to create an output that looks good when viewed in a PowerPoint presentation, consider creating a PNG output with a smaller image size than the default size. In addition, use a style template to make some of the fonts smaller, while rendering that image at 300 DPI. These combined techniques can help to make a graph with text that is more defined and clear.

When determining the best options for generating your graph, you need to take into consideration who the graph is being generated for. Some publications have very specific requirements that need to be met in order to submit. For example, a particular publication might request TIFF images from you, and you can output a TIFF file by using the OUTPUTFMT option.

If you anticipate a lot of back and forth with the people requesting the graph to make minor tweaks such as labels, then you might want to create editable graphics. You can create an editable graph by creating an SGE file which the customer can open with the ODS Graphics Editor.

Some countries use decimal commas instead of decimals. If you need to produce decimal commas on the axes tick values you can use the TICKVALUEFORMAT = NLNUMX.X option in conjunction with the LOCALE system option to produce values with decimal commas.

References

Eslinger, J. (2016). "The Dynamic Duo: ODS Layout and the ODS Destination for PowerPoint." Proceedings of the *SAS Global Forum* 2016 Conference. Las Vegas: Cary, NC: SAS Institute Inc. Retrieved from
https://support.sas.com/resources/papers/proceedings16/SAS5443-2016.pdf

Format My Manuscript. (2020, 07 21). Retrieved from Journal of Clinical Oncology:
https://ascopubs.org/jco/authors/format-manuscript

Heath, D. (2018). "Diving Deep into SAS® ODS Graphics Styles." Proceedings of the Annual Pharmaceutical SAS Users Group Conference. *PharmaSUG* (p. 3). Cary, NC: SAS Institute Inc. Seattle: PharmaSUG. Retrieved from
https://www.pharmasug.org/proceedings/2018/DV/PharmaSUG-2018-DV02.pdf

Matange, S. (2012, March 22). "High resolution graphs." SAS Blogs: Graphically Speaking:
https://blogs.sas.com/content/graphicallyspeaking/2012/03/22/high-quality-graphs/

Matange, S. (2013). *Getting Started with the Graph Template Language in SAS®: Examples, Tips, and Techniques for Creating Custom Graphs*. Cary, NC: SAS Institute Inc.

Matange, S. (2019). "Effective Graphical Representation of Tumor Data." Proceedings of the Annual Pharmaceutical SAS Users Group Conference. Cary, NC: SAS Institute Inc. Philadelphia: PharmaSUG.

SAS Institute Inc. (2020, July 30). *SAS Graph Template Language: User's Guide, Fifth Edition: Creating a Graph That Can be Edited*. Retrieved from SAS® 9.4 and SAS® Viya® 3.5 Programming Documentation:
https://documentation.sas.com/?docsetId=grstatug&docsetTarget=n1tcw0a18pzipcn1xn8roddvkrso.htm&docsetVersion=9.4&locale=en

SAS Institute Inc. (2020, July 27). *National Language Support (NLS), Fifth Edition: LOCALE System Option*. Retrieved from SAS® 9.4 and SAS® Viya® 3.5 Programming Documentation:
https://documentation.sas.com/?docsetId=nlsref&docsetTarget=n1n9bwctsthuqbn1xgipyw5xwujl.htm&docsetVersion=9.4&locale=en

SAS Institute Inc. (2020, August 19). *SAS/GRAPH 9.4: Reference, Fifth Edition: Modifying a Style*. (SAS, Editor) Retrieved from SAS:
https://documentation.sas.com/?docsetId=graphref&docsetTarget=p0qch95nofuklin1k32vhojz1bvq.htm&docsetVersion=9.4&locale=en#p0v767y34gb74bn1qo53mwadh6dm

SAS Institute Inc. (n.d.). SAS® 9 Graph Style Tip Sheet. SAS. Retrieved from SAS:
https://support.sas.com/rnd/app/ODSGraphics/TipSheet_GraphStyle.pdf

SAS Institute Inc. (n.d.). SAS® 9 ODS Tip Sheet. SAS. Retrieved from https://support.sas.com/rnd/base/ods/scratch/ods-tips.pdf

Tabernero, J. et al.. (2018). "Analysis of angiogenesis biomarkers for ramucirumab efficacy in patients with metastatic colorectal cancer from RAISE, a global, randomized, double-blind, phase III study." *Annals of Oncology*, volume 29, no.3, 602-609.

Watson, R. (2020). "What's Your Favorite Color? Controlling the Appearance of a Graph." Proceedings of the *SAS Global Forum* 2020 Conference. Cary, NC: SAS Institute Inc.

Ready to take your SAS® and JMP®skills up a notch?

Be among the first to know about new books,
special events, and exclusive discounts.
support.sas.com/newbooks

Share your expertise. Write a book with SAS.
support.sas.com/publish

Continue your skills development with free online learning.
www.sas.com/free-training

sas.com/books
for additional books and resources.

THE POWER TO KNOW.

www.ingramcontent.com/pod-product-compliance
Lightning Source LLC
LaVergne TN
LVHW080114070326
832902LV00015B/2580